S0-FPH-079

Fission: A > 60	Activation A ≤60
Halogens: Iodine (I) 131 (8 d) **Noble Gas:** Krypton (Kr) 85m (4.4 h) Krypton (Kr) 87·(76.3 m) Xenon (Xe) 133 (5.25 d) **Soluble Metal Ions:** Strontium (Sr) 90 (28.6 y) Cesium (Cs) 137 (30.17 y) Barium (BA) 140 (12.79 d) **Other:** Tritium (H) 3 (12 y)	**Corrosion:** Manganese (Mn) 54 *(312 d)* Cobalt (Co) 58 *(71 d)* **Cobalt (Co) 60 (5.27 y)** Iron (Fe) 59 *(44.5 d)* **Gas:** *Nitrogen (N) 13 (10 m)* Nitrogen (N) 16 (7 s) *Nitrogen (N) 17 (4 s)* *Oxygen (O) 19 (27 s)* Fluorine (F) 18 (110 m) Argon (Ar) 41 (109 m)

1 MeV Gamma	HVL	TVL
Lead	.5"	1.5"
Steel/Iron	1"	3"
Concrete	4"	12"
Water	8"	24"

Air Samples (uCi/cc)	ncpm/(Liters)(2.22E8)
Point Source	$6CEN/d^2$
Half Life	$A = A_o \times 0.5^{t/t_{1/2}}$
$T_{1/2} =$	.693/lambda
Effective Half Life (T_{eff})	$\dfrac{(T_{1/2} \times T_b)}{(T_{1/2} + T_b)}$
Inverse Square Law	$I_1(d_1)^2 = I_2(d_2)^2$

Alpha Level 3 Area: bg/a <300
Alpha Level 2 Area: bg/a= 300-30,000
Alpha Level 1 Area: bg/a >30,000

NUF Cram Notes is a concise NUF study guide from the technician's prospective.

Rennhack has taken the ALARA principal and applied it to a study guide. It's the perfect last minute cram.

© 2015, By Michael D. Rennhack

No part of this publication may be reproduced, stored in a retrieval system or transmitted in any form or by any means without the prior written permission of the publisher.

NUF CRAM NOTES

2016 Edition

Rennhack's Concise Study Guide for the Contract Radiation Protection Technician Nuclear Utilities Fundamentals (NUF) Exam

By Michael D. Rennhack

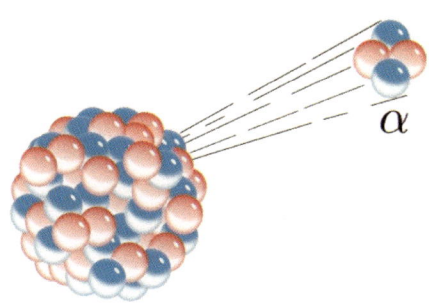

Contents

Introduction 8
About This Book 10
How This Book is organized 11
1 Radiation, Ionization & Interactions with Matter 12
2 Radiation & Contamination Sources 24
3 Radiation Detection & Instrumentation 34
4 Surveys & Job Coverage 54
5 Air Sampling 62
6 ALARA & Shielding 70
7 Decay Modes, Decay Rates, Half-Lives, and the Curie 92
8 Alpha Monitoring & Control 100
NukeWorker Online Practice Test 107
Meter Reading Training 108
Reference Material 120
About The NUF 126
Resume Tips 130
Acceptable Experience & Training for HP/RP Techs 132
Old NEU RP Exam 136
NUF Governing Bodies 138
Other RP Tests Available 140
Issues with NUF 'Facts' 143
History of NukeWorker.com 145
About the Author 151

Introduction

Welcome to NUF Cram Notes, Forth Edition. This book contains concise, highly targeted material, focused on the 52 objectives that the Nuclear Utilities HP/RP Technician Fundamentals (NUF) Exam is based on.

Prior to the publication of 'NUF Cram Notes', technicians were directed to the old NEU study guide as the only available study material. The NEU study guide is a 500 page document that has 85 listed objectives that the original NEU test was based on, but not the exact 52 objectives that the current NUF test is based on. In fact, 50% of the standard NEU study guide, entire chapters, is not testable material. Chapters one "NP-1 Atomic Nature of Mater", two, "NP-2 Nuclear Stability Concepts", and five, "NP-5 Nuclear Reactions" have **no** objectives covered on the test, which means there will be no questions on those chapters. Chapters four "NP-4 Particle Behavior/Gamma Interactions, and chapter six "RP-1 the Biological Effects of Ionizing Radiation" each only have **one** testable objective. To summarize; there are only two test questions covering five out of ten chapters of the original NUF study guide. It is valuable information, however it isn't on the test.

Alpha questions update

In September of 2014, the NUF steering committee added a few more questions about alpha, based on the new EPRI guidance. In addition, they also removed a few questions that had a high failure rate. As a result, there are now 52 questions on the test, of which three of those objectives are on the new alpha material. The new alpha material is in chapter 8.

About This Book

This book is not intended to teach you everything about health physics. It is intended to highlight the 52 objectives the NUF is based on and to review the most critical information pertaining to those objectives. There are already several 500 page books available that explain everything you ever wanted to know about Radiation Protection.

This book is intended to be a quick refresher, for technicians that have already reviewed the full text of the NUF study guide, especially those that have already passed the NUF RP fundamentals test.

The wording and phrasing of the study guide is similar to the wording of the NUF test in order to make the reader familiar not only with the technical content of the test, but also with the manner in which the questions are asked and answered. This is done so that when the actual test is taken it does not feel foreign, but familiar.

How This Book is organized

You will be presented with each of the 52 objectives the current test questions are based on. You will then be given the most critical information pertaining to that objective.

 Pro Tip: Tips on how to remember the critical information will be presented next in this format.

If additional information is available to help you understand or remember the critical information, it will be presented next, and in a grey italicized font. It looks like this. This information is not expected to be on the test and can be skipped.

The easier to digest material is in the beginning of the book, and the math intensive material is at the end. This is done so that your brain won't be numb and exhausted from the math, so that you can focus on the rest of the book. In placing the math at the end, the book allows you to absorb the highest quantity of testable material in the most efficient manner. There are only 4 objectives that require you to do calculations. Reading the book slowly, taking time to memorize as much as possible, you should still be able to finish the book in three hours. No one wants to spend weeks brushing up for a test they have to take every few years when a few hours are more than enough.

Chapter 9; "NukeWorker Online Practice Test", and everything beyond it is supplemental reference material and is not directly addressed on the test, however you may still find it useful.

1 Radiation, Ionization & Interactions with Matter

This is a massive and fundamental section with a great deal of information to cover. However, the NUF only covers two objectives (asks two questions) about it. As a result, **a lot of non-critical information is included in this section, <u>which can be skipped</u>**.

As a reminder, non-critical information looks like this, and can be skipped.

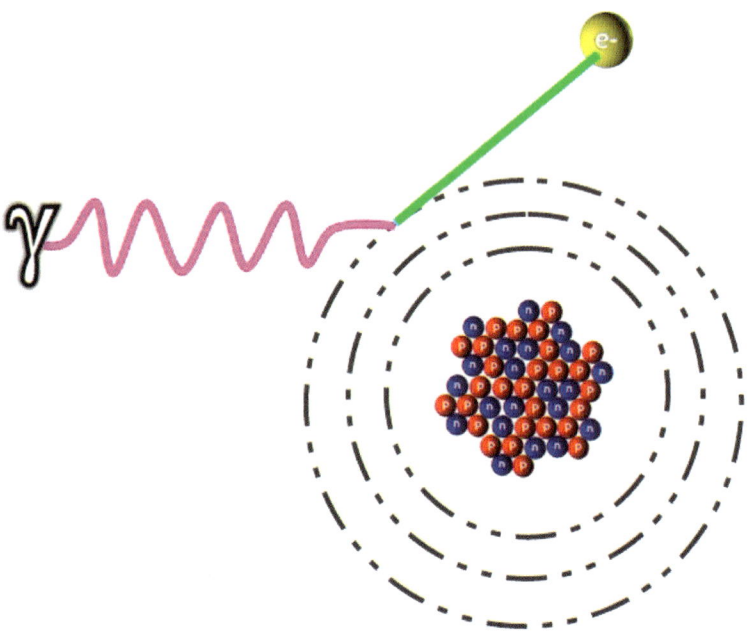

1.1 Describe the various types of radiation, their structure, source, approximate range in air, energy levels, charge and method of interaction / attenuation

Radiation	Structure	Charge	Range in Air	Method of Interaction
Alpha	2 Protons 2 Neutrons	+2	A few Inches	Ionization Excitation
Beta	Similar to Electron	-1	~12 Feet	Ionization Excitation Bremsstrahlung
Gamma	Photon	N/A	Several Hundred Yards	Photoelectric Effect Compton Scatter Pair Production
Neutron	Neutron	N/A	Several Hundred Yards	

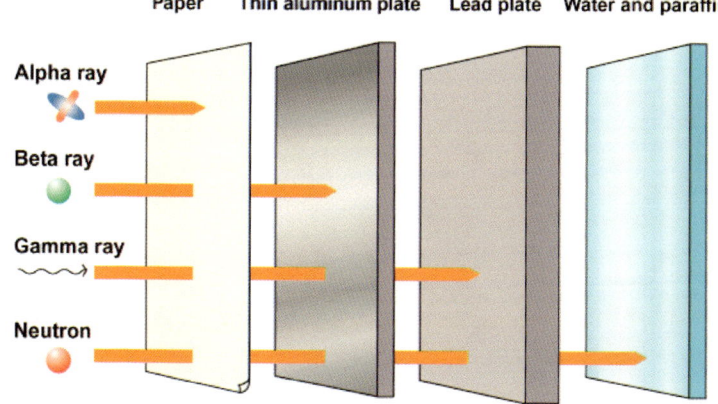

Figure 1 Radiation Penetration

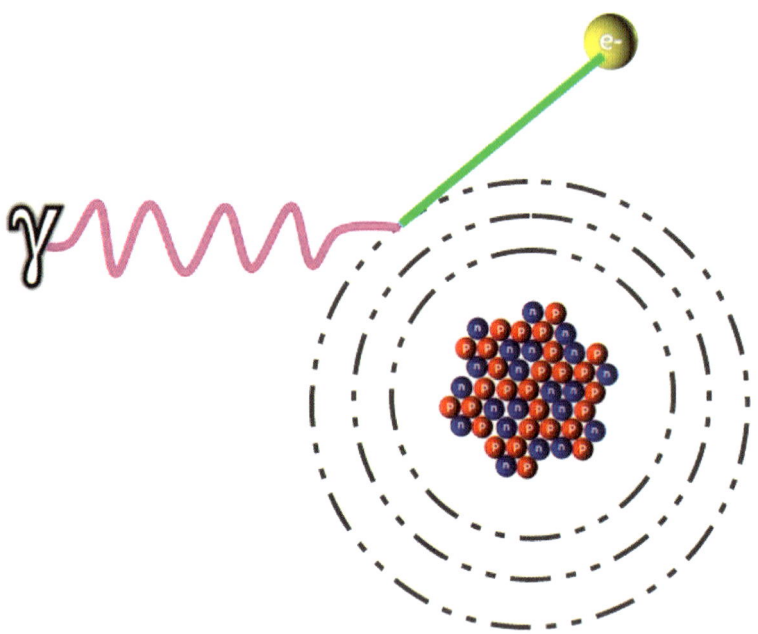

Figure 2 Ionization

Ionization *is any process which results in the removal of a bound electron (negative charge) from an electrically neutral atom by adding enough energy to the electron to overcome its binding energy. This leaves the atom with a net positive charge. The result is the creation of an ion pair made up of the negatively charged electron and the positively charged atom.*

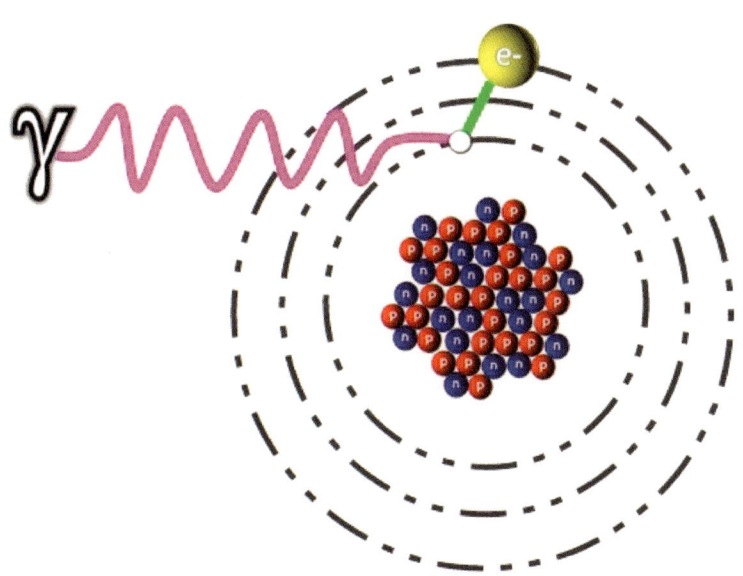

Figure 3 Electron Excitation

Electron Excitation *is any process that adds enough energy to an electron of an atom so that it occupies a higher energy state than its lowest bound energy state. The electron remains bound to the atom. No ions are produced and the atom remains electrically neutral.*

Nuclear Excitation *is any process that adds energy to a nucleon in the nucleus of an atom so that it occupies a higher energy state. The nucleus continues to have the same number of nucleons and can continue in its same chemical environment.*

1.1.1 Alpha

Alpha Particles are Helium Nuclei emitted from the Nucleus of an unstable atom. They consist of 2 protons and 2 neutrons, giving it a mass of 4 AMU.

Due to being the largest sized particle and having a plus two charge it is easiest to attenuate (shield) **Alpha** radiation.

Alpha radiation passing through air would create the highest specific ionization value.

The approximate range (mean free path) in air for a 7 MeV alpha particle is approximately **2 inches.**

Alpha is long lived compared to beta-gamma, as beta-gamma levels go down, the alpha hazard increases. Alpha is primarily an internal hazard so PAS (personal air samplers) are issued as dosimetry devices to measure the intake of activity for work in Level III (high hazard) areas. The internal dose from alpha is 1,000-10,000 times the dose from the same beta-gamma activity.

*The new alpha material is found in chapter 8.

1.1.2 Beta

A Beta particle is an electron (1/1836 mass of a proton) emitted from the <u>nucleus</u> of a radioactive nuclide.

The approximate range (mean free path) for a 1 MeV Beta particle in air is approximately **12 feet**. However the range is dependent on the **energy of the beta particle.**

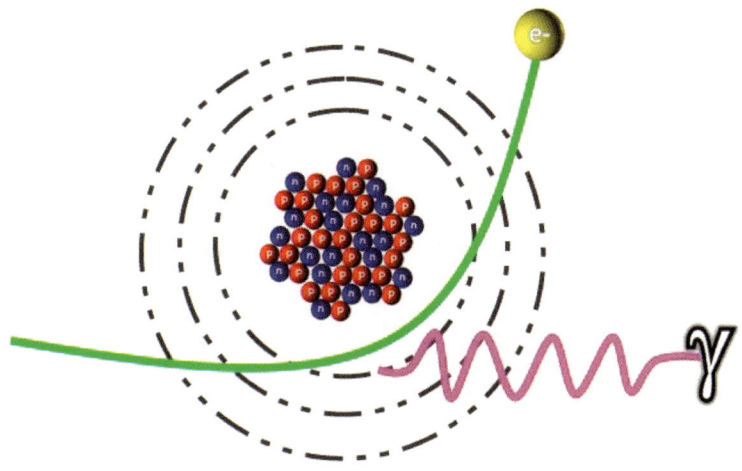

Figure 4 Bremsstrahlung

Bremsstrahlung, a form of Beta interaction**,** is the energy emitted during deceleration of a charged particle. Bremsstrahlung is caused by a decrease in energy of a Beta particle from coming in close proximity to a <u>heavy</u> nucleus. This is why lead is a bad shield for beta radiation.

1.1.3 Gamma

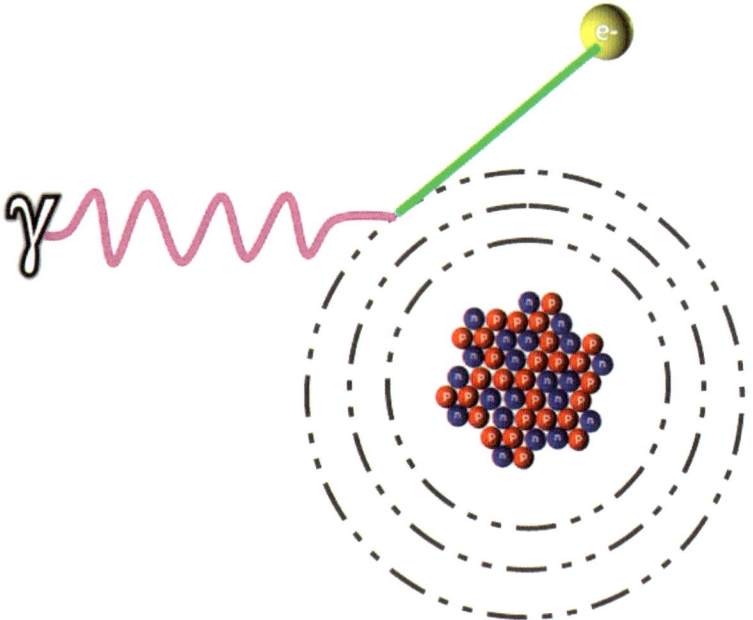

Figure 5 Photoelectric Effect

Photoelectric Effect is the process where an atom adsorbs a Gamma ray and then ejects an orbiting electron.

The photoelectric effect is only significant for photon energies less than 1 MeV.

Compton Scatter: In the shielding of Gamma rays, the dose buildup factor accounts primarily for the dose contribution due to **Compton Scattered** photons.

In Compton scattering there is a partial energy loss for the incoming photon. The photon interacts with an orbital electron of some atom and only part of the energy is transferred to the electron. Compton scattering is the dominant interaction for most materials for photon energies between 200 keV and 5 MeV.

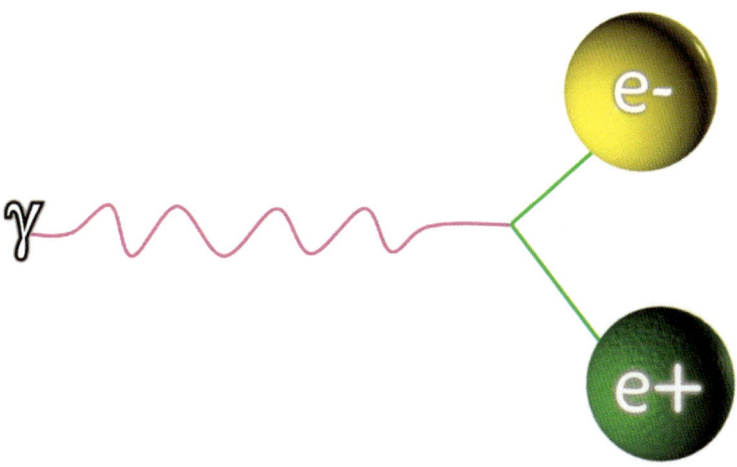

Figure 6 Pair Production

Pair Production is the process where a Gamma ray (photon) is transformed into a Positron and Electron.

In pair production, a gamma photon in the vicinity of a nucleus 'disappears', and in its place appears a pair of electrons: one negatively charged and one positively charged.

This conversion of energy to mass only occurs in the presence of a strong electric field, which are found near the nucleus of atoms and are stronger for high Z materials. Additionally, it is impossible unless the gamma ray possesses greater than 1.022 MeV of energy to make up the mass of the particles. It does not become important until >2 MeV to give kinetic energy to the particles.

1.1.4 Neutron

The neutron has a mass number of 1 and no charge. Because it has no charge the neutron can penetrate relatively easily into a nucleus. Free unbound neutrons are unstable (radioactive) and disintegrate by beta emission with a half-life of approximately 10.6 minutes. The resultant decay product is a proton.

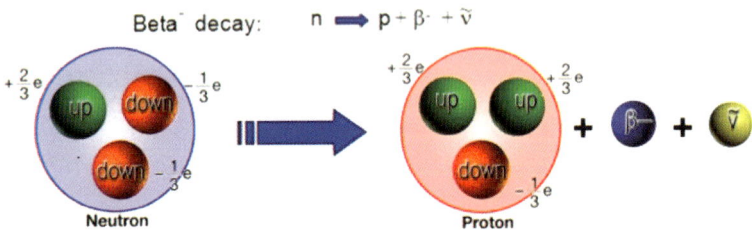

Figure 7 Neutron Decay

Neutrons are made up of two down quarks, and an up quark. During the decay process a down quark changes into an up quark, leading to the emission of an electron and an antineutrino.

1.2 Describe the characteristics of Alpha, Beta and Positron decay processes

1.2.1 Alpha decay

Alpha decay may occur in nuclei having a mass number greater than 200 (sometimes stated as greater than 210).

For example, if an atom of **Po-210** undergoes an Alpha decay, the daughter product is **Pb-206**.

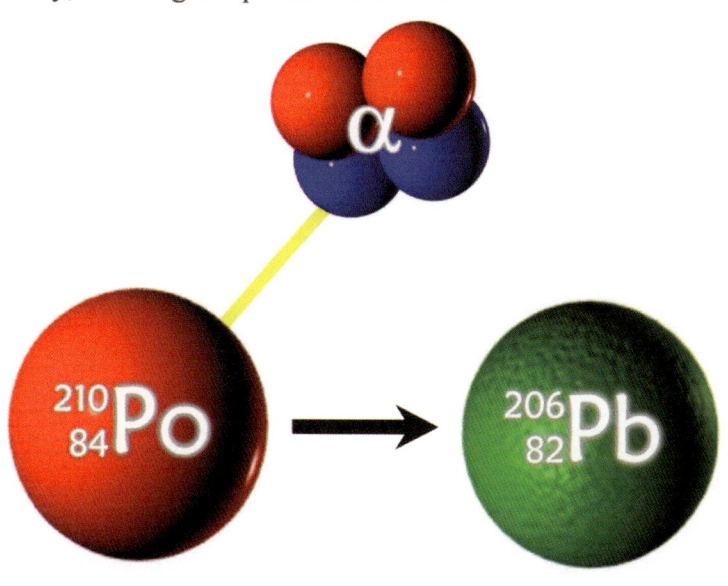

Figure 8 Po-210 Undergoing Alpha decay

1.2.2 Beta minus Decay

Beta minus Decay: Emission of a Beta particle results from the transformation of a Neutron, in the nucleus, into a Proton and an Electron.

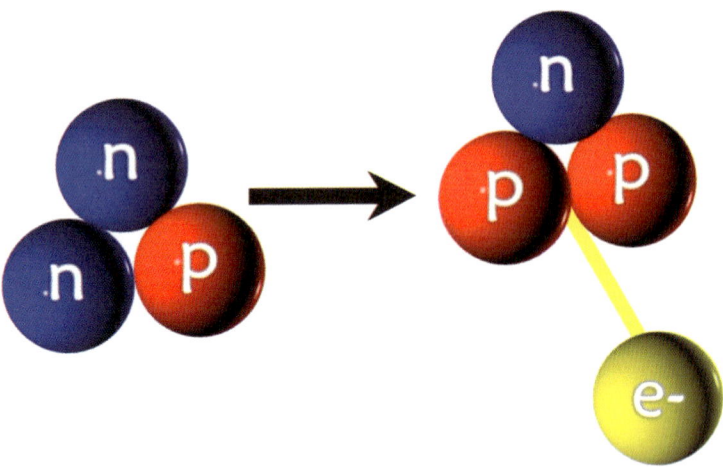

Figure 9 Beta minus Decay

1.2.3 Positron (Beta Plus) Decay

Positron (Beta Plus) Decay: In instances where the n/p ratio is below the stability curve, a nucleus will decay in a manner that will increase the ratio of neutrons to protons. One of the decay processes a nucleus will undergo to achieve this is **Positron (Beta Plus) Decay**.

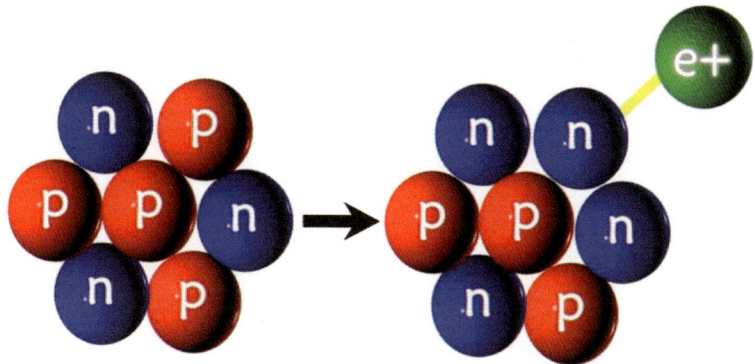

Figure 10 Positron (Beta Plus) Decay

2 Radiation & Contamination Sources

There are 10 objectives the NUF covers from this category, which is 20% of your score. It is the second largest category and should be studied closely.

2.1 Identify methods of plant radiation source production

Fission products and Neutron activation of reactor coolant corrosion products (Activation products) are the two major sources of plant radiation exposure/source production during shutdown maintenance and normal operations.

2.1.1 Fission product

Fission products are the most abundant during reactor/power plant operation (because most of them have short half-lives).

Fission product activity in reactor coolant is a major category of plant radiation source production during shutdown maintenance.

2.1.2 Activation Product

Neutron activation of reactor coolant corrosion products is a method (major category) of plant radiation source production & radiation exposure during shutdown maintenance.

2.2 Describe three methods that result in fission products in plant systems

Fission product radioactivity in reactor coolant/plant systems can result from:

- Fuel rod/element cladding failure
- Fuel element defects
- Leaking fuel pins
- Tramp Uranium on the outside of the fuel cladding
- Fuel Cladding material contaminated with Uranium impurities within the fuel claddings

 Pro Tip: Stellite valve seating surfaces is NOT a source of fission products in the plant systems (it's a source of activation products.)

2.3 Explain the reason for monitoring fission product activity in the coolant

Monitoring reactor coolant **Fission product** activity provides the most accurate method to determine if there has been a fuel element cladding failure on a fuel rod during normal operations. Because an **increase in Reactor Coolant fission product activity** could be due to **fuel rod cladding damage/failure.**

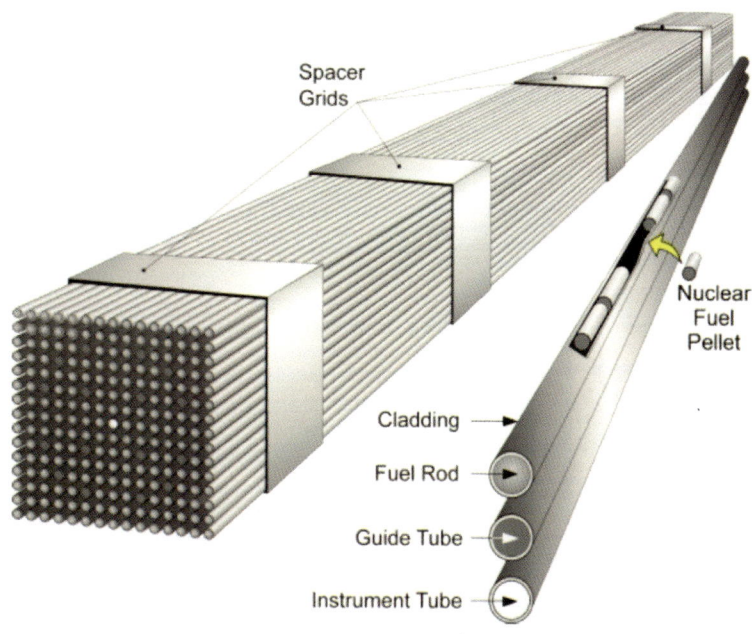

2.4 Identify the common fission products

Fission Products can typically be identified in a line up with Activation products because most of them have an **Atomic number greater than 60**, for example:

Halogen Fission Products:
- Iodine (I) 131 (8.04 days)

Noble Gas Fission Products:
- Krypton (Kr) 85m (4.38 hours)
- Krypton (Kr) 87 (76.3 minutes)
- Xenon (Xe) 133 (5.25 days)
- *Xenon (Xe) 131m (11.84 days)*

Soluble Metal Ion Fission Products:
- **Strontium (Sr) 90 (28.6 years)**
- **Cesium (Cs) 137 (30.17 years)**
- Barium (BA) 140 (12.79 days)

Other Fission Products include:
- **Tritium (H) 3 (12 years)**

2.5 Recognize the methods of removal or escape of halogens, noble gases and soluble metal ions from the reactor coolant

Halogens such as io<u>d</u>i<u>n</u>e are normally removed from the reactor coolant by using purification <u>ion</u> exchange.

Pro Tip: One way to remember that Iodine is removed via Ion exchange is the letters i.o.n in Iodine.

Noble <u>gas</u>es such as Krypton and Xenon are a major source of activity in a PWR & BWR main steam line and are normally removed from the reactor coolant by venting to the <u>Gaseous</u> Radwaste System, the environment, and using a <u>Degasifier</u>.

Pro Tip: It should be easy to remember that <u>gas</u>es are removed via de<u>gas</u>ifier, or <u>gas</u>eous Radwaste system.

Soluble metal <u>ions</u> such as **Cesium**, **Barium**, and **Strontium** are normally removed from the reactor coolant by using purification <u>ion</u> exchange.

Pro Tip: It should be easy to remember that ions (Soluble metal ions) are removed via ion exchange.

2.6 Identify activated corrosion products found in a light water reactor

Co-60, an **activated corrosion product**, decays producing two high energy-gammas 100% of the time and is a major contributor of reactor coolant activity/dose rates in a nuclear power plant.

Ni-58 is a major component of stainless steel and **Inconel** and when activated produces Co-58 which is an **activated corrosion product** and a major contributor to the reactor coolant activity level/dose Rates.

Activation products can typically be identified in a line up with Fission products because most of them have an **Atomic number ≤60**, for example:

Activated Corrosion:
- Manganese (Mn) 54 *(312.03 days)*
- Cobalt (Co) 58 *(70.86 days)*
- **Cobalt (Co) 60 (5.27 years)**
- Iron (Fe) 59 *(44.5 days)*

2.7 Describe the radiological impact of the use of Stellite in primary system components

Stellite material is used in reactor parts and components in the reactor coolant system. Stellite valve seating surfaces is a source of <u>activation products</u> in the plant systems.

The reason why **Stellite** is a radiological hazard in a nuclear environment is because Stellite contains a high percentage of Cobalt (Co) 59 which becomes highly radioactive Co-60 when activated via Neutron bombardment.

$$^{59}_{27}Co + ^{1}_{0}n \longrightarrow ^{60}_{27}Co$$

$$^{60}_{27}Co \longrightarrow ^{60}_{28}Ni + ^{0}_{+1}\beta + \gamma$$

2.8 Identify sources of activity in a Boiling Water Reactor (BWR) steam line during operation

Nitrogen (N) 16 is the primary source of radiation in a **Boiling Water Reactor** (BWR) main steam line during normal operations; however, it is not a problem immediately after shutdown (due to 7.14 second half life).

Sufficient delay time must be taken in the design of sampling lines to assure decay to negligible levels before the coolant reaches the sample point for N-16.

Any gaseous activity in a BWR will go with the steam and eventually pass out of the system via the air ejector. The Nitrogen-16 nuclide emits gamma rays with energies of 6.13 and 7.11 MeV upon decay and is the major source of activity through the main coolant system of both BWRs and PWRs.

BWRs operating under Hydrogen Water Chemistry (HWC) conditions tend to have higher Main Steam Line dose rates due to the production of ammonia in the Reactor and it's carryover in the steam. This ammonia carryover consists of extra Nitrogen-13 & N-16 which may triple the dose rates over non-HWC environments.

Activated Gas:
- *Nitrogen (N) 13 (9.96 minutes)*
- Nitrogen (N) 16 (7.14 seconds)
- *Nitrogen (N) 17 (4.17 seconds)*
- *Oxygen (O) 19 (26.9 seconds)*
- Fluorine (F) 18 *(Soluble)* (110.0 minutes)
- Argon (Ar) 41 (109.34 minutes)

The gaseous activity in the reactor coolant is due to activation of Oxygen or Argon-40, both are found in air.

2.9 Identify reactor coolant gaseous activity contributors and their effect of post shutdown dose rates

One contributor to reactor coolant gaseous activity during power operations is **Tritium, a fission product.**

Tritium is a low-yield fission product but has a long Half-life of 12 years. It easily diffuses through Zircaloy fuel cladding and combines with oxygen in the reactor coolant to form tritiated water, which follows the same chemistry as hydrogen. Tritium is not removed by the various purification systems in the plant and will accumulate as the plant operates.

A contributor shortly after the reactor is shutdown in a PWR is **Xenon** (Xe) 133 (5.25 day half-life).

Argon (Ar) 41 contributes to post shutdown dose rates 5 minutes after shutdown (109 minute half-life).

Some of the gaseous activity in the reactor coolant is due to Argon-41, which occurs from the activation of Argon-40, argon being a component of air. If a high Argon-41 activity level does occur, it is usually due to an introduction of air into the reactor coolant system via makeup water.

3 Radiation Detection & Instrumentation

There are 14 objectives covered on the NUF from this category, which is 28% of your score. It is the largest category and the most important. This is the section that most people struggle with, so plan to spend extra time on this material.

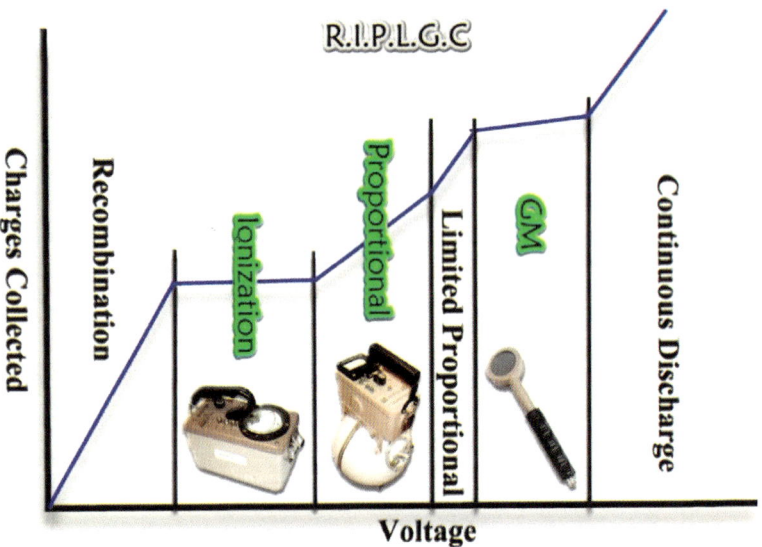

Figure 12 Six Region Gas Amplification Curve

3.1 Explain how ionization can be used to detect radiation

Gas filled radiation detectors work on the principle that radiation ionizes (creates **ion pairs**) the fill gas in the detector material releasing electrons which are collected at a positive electrode which results in a current flow that can be measured by the electrical circuitry, which is converted into a meter reading used to indicate the amount of radiation present.

As radiation passes through the chamber of a gas filled detector to be detected it may interact with the chamber walls, the fill gas, and with the material within the chamber.

There are several factors which affect the number of ion pairs that are produced in a gas filled detector. These factors include:

- Type and energy of the radiation
- Type and density of the fill gas
- Size of the detector
- Voltage applied to the detector

3.2 Describe the basic operation for a typical Gas-filed detector circuit

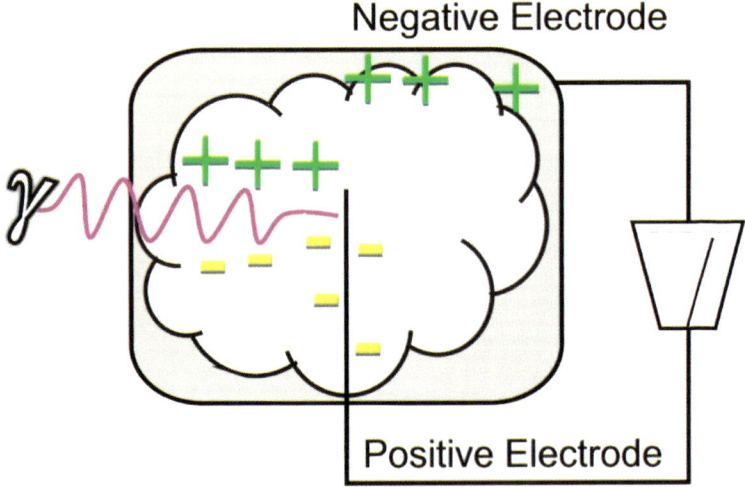

Figure 11 Gas Filed Detector

In a typical Gas Filled Detector, the battery (power supply) in the circuit effectively creates poles (or electrodes) in the detector. The center wire or rod is the positive electrode and the walls of the chamber are the negative electrode.

Gas amplification is the production of secondary ionization by the initial ions produced by the radiation in a gas filled detector.

3.3 Describe the basic operating characteristics of the Ionization Chamber

In a gas filled detector that operates in the **Ionization Chamber Region** of the gas amplification curve, every ion pair produced is collected on the electrodes (operating principle). There is no recombination and no gas amplification.

In the operation of an **Ion Chamber** the signal produced by ion pairs in the chamber stays the same as the applied voltage is increased or decreased in the ion chamber region. The number of ion pairs collected in the **ionization region** does not vary with voltage.

20 - 25% is the approximate beta efficiency range (in percentage) for **Ion Chambers**.

3.4 Describe the basic operating characteristics of a Proportional Counter

In a gas filled detector that operates in the **Proportional Region** of the Six Region (gas amplification) curve the number of ion pairs collected on the electrodes is greater than the number of ion pairs produced by the radiation due to gas amplification.

In the operation of a **Proportional counter** the pulse size produced by ion pairs in the chamber increases as the applied voltage increases in the **Proportional Region**.

As long as the **voltage remains constant**, the number of ion pairs collected in the proportional region is proportional to the number of ion pairs originally produced in the detector by radiation.

3.5 Explain how a Proportional Counter can discriminate between different types of radiation

The Proportional counter discriminates between types and energies of radiation by using a pulse-height discriminator; the pulse size will be larger for alpha than for beta particles.

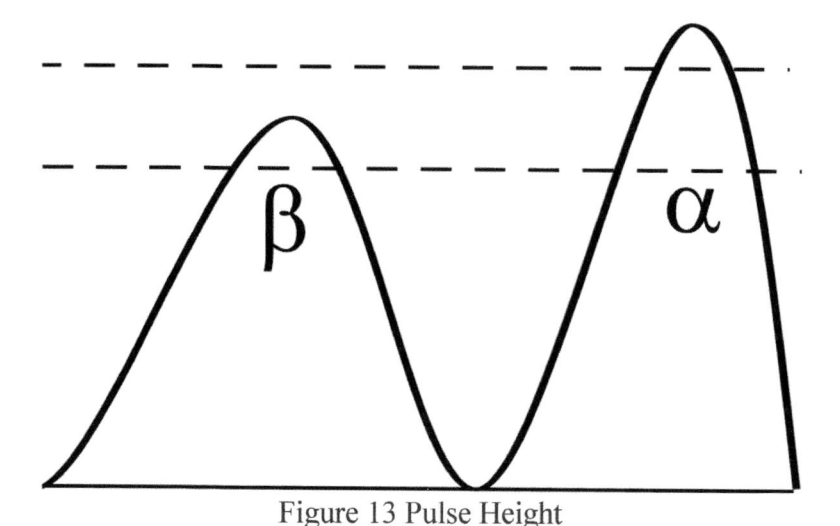

Figure 13 Pulse Height

3.6 Explain how a Proportional Counter can be used to measure Neutron dose rates

In a Proportional counter that is used to measure Neutron dose rates in the field, thermal neutrons are absorbed by the Boron (B) 10 in the fill gas and detector lining. This produces Lithium (Li) 7 and an alpha particle which then causes ionization in the gas producing ion pairs that cause current to flow.

$$n + B10 = Li7 + a$$

3.7 Describe the basic operating characteristics of a Geiger-Mueller (GM) detector

In a gas filled detector that operates in the Geiger-Mueller (GM) Region of the gas amplification curve, every ionizing event produces so many secondary ions that a very large pulse is produced.

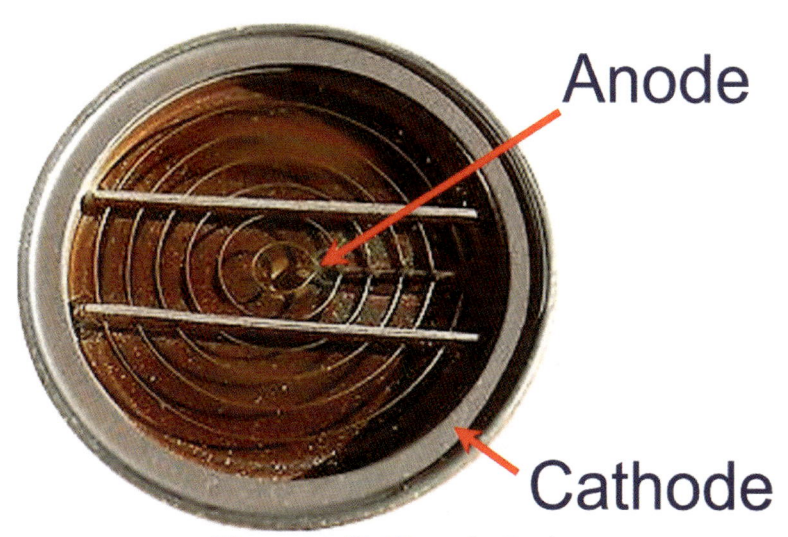

Figure 14 GM Pancake Probe

A quench gas is used in a GM detector to prevent re-triggering Townshend avalanche.

Resolving time in a GM detector is defined as the time needed to clear away the avalanche of ion pairs. Dead time for GM detectors is the period of time in which a second pulse will not be detected.

3.8 Explain the basic operation of a Scintillation detector

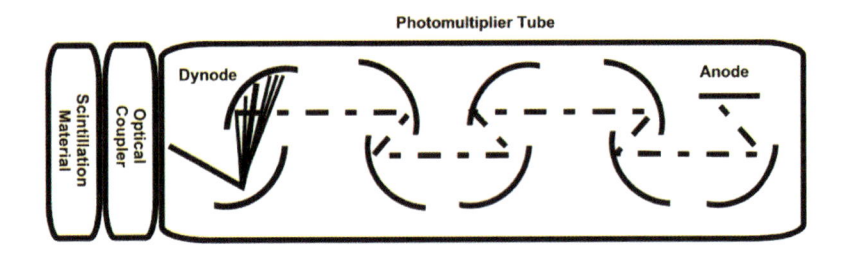

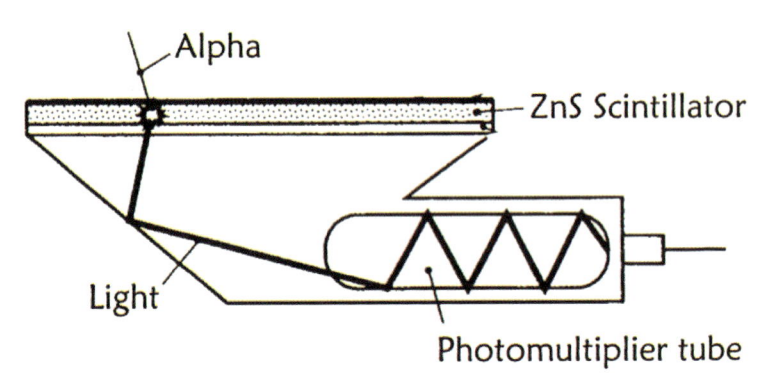

Figure 15 Scintillation Detector

In a **Scintillation Detector**, when radiation interacts with the scintillation material visible light is given off, the optical coupler (**light pipe**) functions to direct the light produced into the **photomultiplier tube**. The function of a **dynode** in a photomultiplier tube is to attract and multiply electrons from the previous dynode.

Figure 16 'Canned' Scintillation Detector

Crystals in the detector are **canned** in metal, usually aluminum, except for the end, which is attached to the photomultiplier tube.

3.9 Explain why Scintillation detectors are more sensitive than Gas-filled detectors

Scintillation detectors are more sensitive than gas filled detectors because the photo-multiplication process allows a small level of radiation to produce a large pulse that is proportional to the initial level of radiation.

A Gas-Filled Detector uses a material with a lower density than a Scintillation Detector. The difference in density makes the scintillation detector more sensitive to lower activities of the radiation being measured because there is more material to interact with.

3.10 Explain the basic operation of a Semi-conductor [detector]

In a **Semiconductor Detector**, the center region acts as the sensitive section of the detector where there is an interaction of positive and negative charges. The size of the pulse is directly proportional to the number of electrons collected.

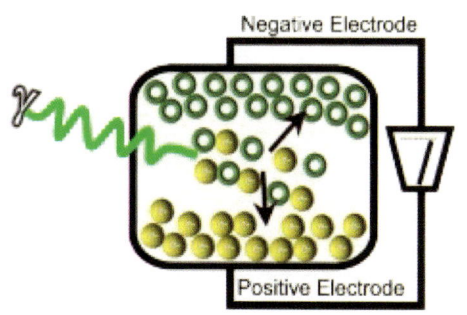

Figure 17 Semiconductor Detector

The theory of how a **Germanium Crystal (Semiconductor) Detector** works is that incident radiation interacts with the atoms in the crystal structure. The interaction converts the energy of the radiation into the flow of free electrons toward a positive electrode and movement of "holes" toward a negative electrode, creating a current flow which results in a measurable pulse.

3.11 Relate the types of radiation detectors to typical applications in the plant

3.11.1 Ion Chamber

A typical use of an Ion Chamber instrument is to measure beta and gamma dose rates in the field.

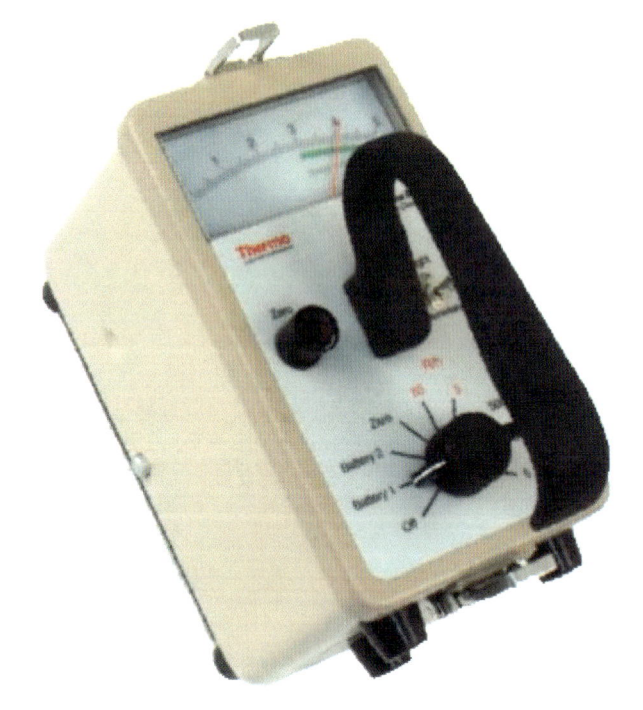

Figure 18 RO-20 Ion Chamber

3.11.2 Proportional Counter

A typical use of a Proportional Counter is to measure Neutron dose rates in the field.

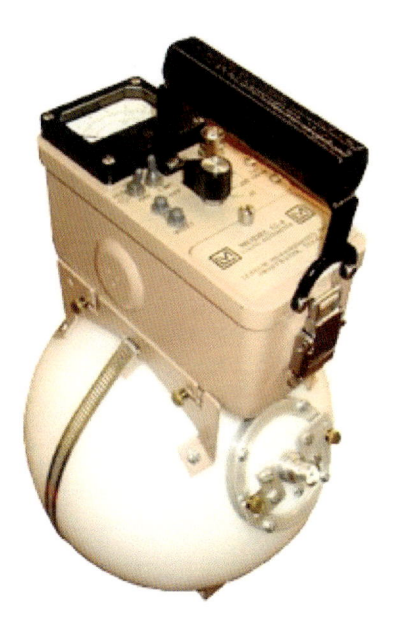

Figure 19 'Rem Ball' Proportional Counter

3.11.3 Geiger-Mueller (GM)

A typical use of a pancake GM instrument is to measure low levels of contamination.

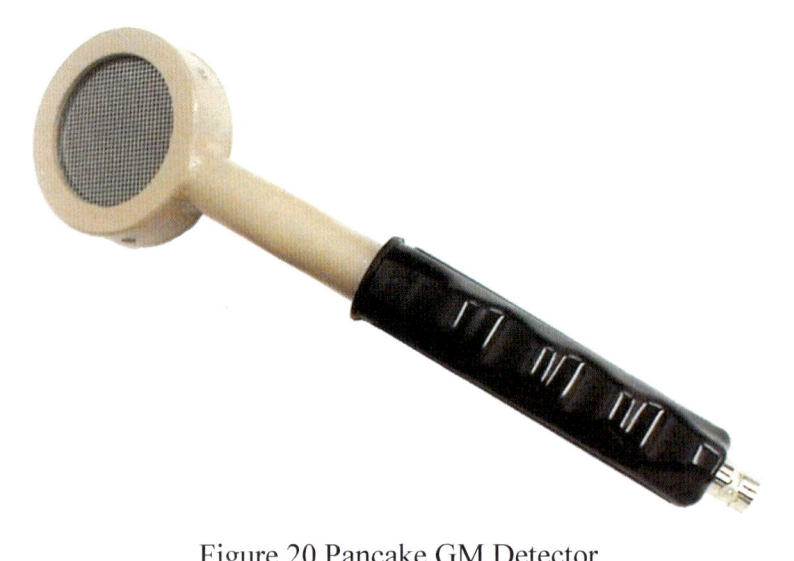

Figure 20 Pancake GM Detector

3.11.4 Scintillation and Semiconductor

Typical uses of scintillation and semiconductor detectors are Radiochemistry and Health Physics counting lab applications.

Figure 21 L2929 Scintillation Scaler

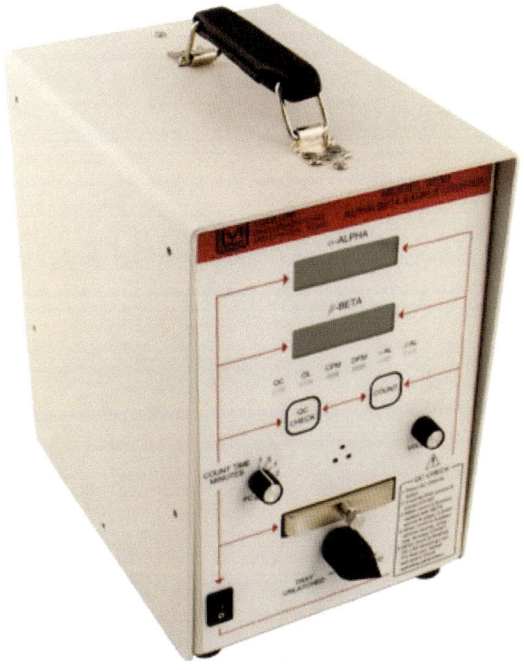

Figure 22 L3030 Semiconductor Scaler

3.12 Define Minimum Detectable Activity (MDA), Minimum Detectable Counts (MDC), and Lower Limit of Detection (LLD) using the terminology of radiological counting

Minimum Detectable Activity (MDA) is defined as the minimum activity present in a sample that may be detected by a particular instrument.

Selecting a counting system with higher efficiency will lower your MDA.

If a system's MDA is not low enough—either because of a low efficiency or high background—it can be improved. This can be done by increasing the counting time used for the measurement, or decreasing the background, or both.

$$MDA = \frac{3 + (4.65 + \sqrt{Bkg})}{Eff}$$

Above is a sample MDA Calculation for a 2929 assuming 10 min Background and a 1 min sample.

50

Lower Limit of Detection (LLD) is defined as the smallest amount of activity that will yield a net count for which there is a confidence at a pre-determined level that activity is present.

The LLD (L_D) for a detector is dependent upon:
- Length of count time
- Detector efficiency
- Detector background

Pro Tip: The LLD (L_D) for a detector is not dependent upon the sample volume.

Minimum Detectable Counts (MDC) is the LLD defined in terms of a count rate since the background counts is time dependent. MDC is that count rate which indicates the presence of activity with a probability of x and with only a (1 - x) probability of falsely concluding its presence.

$$MDC = Bkg + 3 \times (4.65 \times \sqrt{Bkg})$$

Above is a sample MDC Calculation for a 2929 assuming 10 min Background and a 1 min sample

3.13 Describe methods used to assess Beta dose

An individual's beta dose can be determined by multiplying their stay time by the beta dose rate measured by correcting the difference of an unshielded ion-chamber meter reading and a shielded reading.

The Shielding designed into a TLD allows beta dose to be determined by calculating Gamma (Beta/Gamma) dose.

3.14 Describe the advantages and disadvantages of a Thermo Luminescent Dosimeter (TLD)

TLDs Advantages:
- Size: TLD chips are so small that they can be taped to the fingers to measure exposure to the extremities without interfering with work.
- The TLD is generally more sensitive than a film badge, **more accurate in the low mR range**, and able to provide a better overall indication of the total beta/gamma dose received.
- TLD chip can be **reused** after it is read.
- TLD is not as sensitive to moisture as is the film badge, so data would not be lost if the TLD became wet.

TLD Disadvantages:
- Cost
- Loss of dose/no permanent record

Figure 23 TLD

4 Surveys & Job Coverage

The NUF has seven objectives in this category, and they are relatively straight forward.

4.1 List three (3) valid reasons why radiation surveys are performed

Valid reasons for performing a radiation survey include:
- To locate, identify and characterize radiological hazards such as determining area dose rates
- To determine or verify required radiological postings
- Completing REPs/ RWPs

4.2 Explain the importance of a representative radiological surveillance

It is important to obtain **representative radiation surveys** in order to determine appropriate work practices.

Pro Tip: 'Estimating the loose contamination re-suspension factors' is NOT a component of a representative radiological survey.

A breathing zone **airborne contamination** sample is not a component of a representative pre-job radiological survey, (because the person isn't breathing the air yet – it's a **General Area** Air Sample before the guy gets there).

4.3 List three (3) checks made on a survey instrument prior to use

- Calibration date
- Check battery
- Physical Condition
- Source/response Check

4.4 Describe proper practices for performing radiological surveys

When performing a **direct frisk** for contamination, the frisker probe should be kept within **one-half inch** of the material being surveyed and should be moved no faster than one to **two inches per second**.

A smear survey for loose contamination should cover an area of 100^2 **cm**. Loose contamination survey locations are identified on a survey map by placing the **smear number inside a circle**.

A practical method for reducing the potential for **cross-contamination** of smears taken in a contaminated area is for each smear to be placed in an envelope or inside a folded piece of paper and handled carefully.

4.5 Recognize situations in and around contaminated areas that may require increased survey frequency or special surveys

Situations in and around contamination areas that may require **increased survey frequency** or **special surveys** include:
- Hot particles
- Grinding or welding
- Moving radioactive waste
- Removing eddy current probes from the steam generator.

The frequency of airborne sampling may be increased due to increased work activity in a contamination area.

Job Coverage Surveys requires a more thorough contamination survey.

4.6 Describe situations when Stop Work Authority can be exercised.

The following is a list of valid situations where **Stop Work Authority** should be exercised:

- Increasing dose rates
- Unsafe conditions in the work area
- Prescribed ALARA controls not in place
- Rising contamination levels
- Suspected uptake based on contaminated wound
- Alpha levels not covered in RWP/ALARA planning documents or not discussed during the ALARA and pre-job briefings
- When in-progress survey results (i.e. contamination swipes or air samples), change the initial alpha level to a higher level (such as from Level I to Level II)

Pro Tip: A worker reporting to a Control Point with lost dosimetry would *NOT* typically cause you to exercise "Stop Work Authority" (you would send that worker out of the RCA, but not stop the entire job).

59

4.7 List radiological reasons for decontamination

Good reasons for performing decontamination:

- Reduces chances of ingestion
- Reduces the need for protective clothing
- Aids in exposure reduction
- Minimizes airborne activity levels

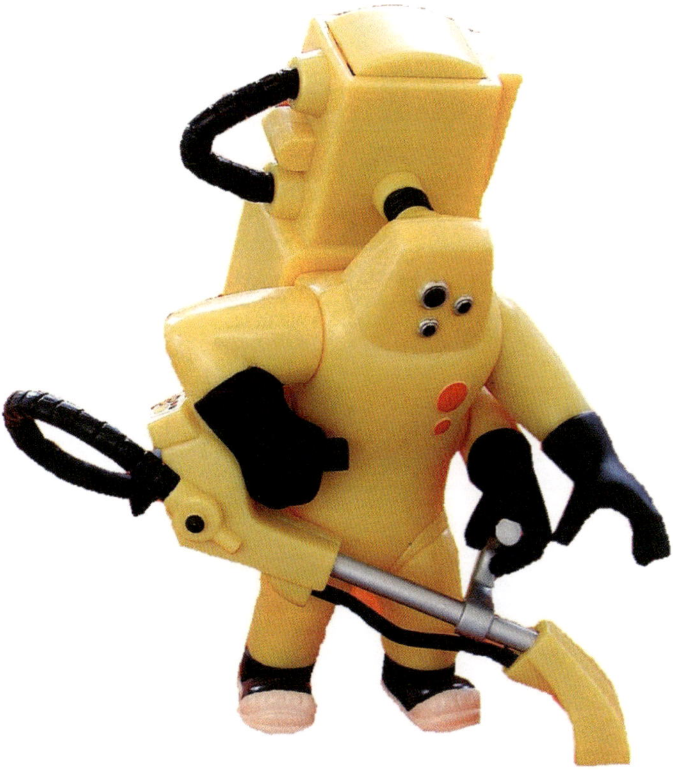

5 Air Sampling

The NUF covers five objectives in this category, one of which will be an Air Sample calculation.

5.1 List potential sources of airborne contamination

Potential sources of radioactive airborne contamination include:

- Removal of HEPA filters from a ventilation unit has the greatest potential for creating an airborne radioactivity area.
- Failure of a HEPA filter
- Welding on radioactive components or systems
- Leak or venting of reactor primary coolant system
- Breach of a containment system (i.e., glove box, tent, etc.)
- Suspension of loose surface contamination
- Radioactive spills
- Grinding and cutting on primary systems can expose alpha that was shielded by an oxide layer, so initial results of low alpha are not always indicative of the final hazard. Periodically re-sample smears and air for alpha.

5.2 Describe some common causes of air sampling error

Sources of **air sampling** errors may stem from:

- Inaccurate sample volumes/count times to detect 0.3 DAC
- Counting instrumentation errors
- Sampler collection efficiency
- Lack of representativeness
- Prevent filter loading due to dust or debris
- Radon (short-lived activity) compensation

5.3 Describe various sampling media

Section 5.3 has been merged with section 5.4 below. There will be two questions from the information in section 5.4

5.4 Describe a method of sampling for each of the following contaminants: Particulates, Iodine, Noble gas and Tritium

Particulates are collected on some type of filter paper such as paper, fiberglass, or cellulose.

Activated charcoal cartridge or **Silver Zeolite** filter media is used for **Iodine** air sampling.

When collecting an air sample, (Noble) **gasses** are collected in a sample chamber that actually contains a small volume of the atmosphere.

A typical method for collecting **Tritium** air samples is passing air through water and counting with **liquid scintillation**.

5.5 Explain the significance of sampling in the breathing zone

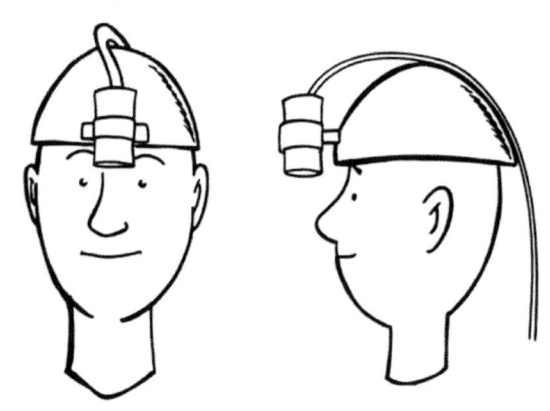

Breathing zone air sampling is taken to assess airborne conditions in the work area and can be described as a sample which represents the airborne concentration the workers are breathing. They are drawn from a point or series of points within the workers immediate work area.

An advantage to taking a breathing zone air sample using a lapel air sampler is that it provides a sample of what was actually breathed. It is always the best location to place the air sample head to obtain a representative sample.

5.6 Calculate the airborne concentration in uCi/cc when given sample and background count rate, volume and counter efficiency

ncpm = gross cpm − background
dpm = ncpm/effenciency
uCi = dpm/2.22E6
Liters = LPM x minutes
cc = Liters/1,000

$$\frac{\text{(gross cpm - background)}}{(LPM)(min.)(1000)(eff)(2.22E6)}$$

That formula can be simplified since all of the questions on the test use 10% efficiency.

$$\frac{(ncpm)}{(LPM)(min.)(2.22E8)}$$

That formula can be simplified further, if the sample volume is listed in liters instead of LPM and duration:

$$\frac{(ncpm)}{(Liters)(2.22E8)}$$

Pro Tip: Divide the net counts per minute by everything.

Practice:
You take an air sample; the flow rate on the air sampler is 2 LPM (Liters per minute), and you run the sample for 10 minutes. You count the sample; the instruments background is 15 cpm (counts per minute), the counter efficiency is 10%, and the air sample's activity is 325 cpm.

Flow: 2 LPM
Duration: 10 min
Sample: 325 cpm
Background: 15 cpm
Instrument efficiency: 10%

(Sample - bkg)/(LPM)(min.)(2.22E8)
(325 counts − 15 bkg)/(2 LPM)(10 min.)(2.22E8)
(310)/(2)(10)(2.22E8)
(155)/(10)(2.22E8)
(15.5)/(2.22E8)
= 6.98E-8 uCi/cc

Practice:
Volume: 500 Liters
Sample: 200 cpm
Background: 80 cpm
Instrument efficiency: 10%

Since we are given total liters, instead of LPM and duration, the math is one step shorter.

(Sample - bkg)/(Liters)(2.22E8) = uCi/cc
(200 counts – 80 bkg)/(500 Liters)(2.22E8)
(120)/(500)(2.22E8) = uCi/cc
(0.24)/(2.22E8) = uCi/cc
= 1.08E-9 uCi/cc

Practice:
Volume: 500 Liters
Sample: 300 cpm
Background: 50 cpm
Instrument efficiency: 10%

(sample - bkgd)/(Liters)(2.22E8) = uCi/cc
(300 gross – 50 bkgd)/(500 Liters)(2.22E8)
(250)/(500)(2.22E8) = uCi/cc
(0.5)/(2.22E8) = uCi/cc
= 2.25E-9 uCi/cc

6 ALARA & Shielding

The NUF covers six objectives in this category, two of which will be calculations.

6.1 Describe methods used to keep radiation exposure As Low as Reasonably Achievable (ALARA)

Good ALARA techniques are:
- **Time:** Minimize time in the work area
- **Distance:** Maximize distance
- **Shielding:** Use temporary shielding
- **Source Removal**: Decontamination of the work area/components.

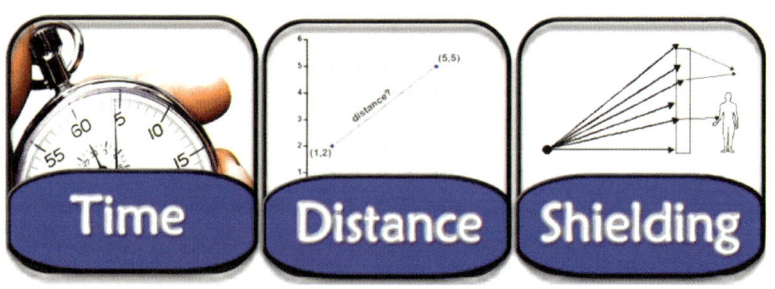

Figure 25 ALARA

To determine when a job meets the ALARA concept, make sure to "Follow the ALARA Program determining if all Administrative, Engineering, Operational and Radiological controls and procedures are in place and being followed."

6.2 Define the following terms: Half Value Layer (HVL), Tenth Value Layer (TVL), and Bremsstrahlung

Half Value Layer (HVL) (or 'thicknesses') is defined as the thickness of a given material required to reduce the photon intensity to one-half the initial value assuming **no** buildup.

Pro Tip: Remember 'assuming **no** buildup'.

Tenth Value Layer (TVL) is the thickness required to reduce the photon intensity by $1/10^{th}$ the initial value.

For example, the amount of shielding required to reduce a 2.5 R/hr Gamma source to 2.5 mR/hr is 3 tenth-value layers.

$$2,500\ mR/hr \xrightarrow{\text{1st TVL}} 250 \xrightarrow{\text{2nd TVL}} 25 \xrightarrow{\text{3rd TVL}} 2.5\ mR/hr$$

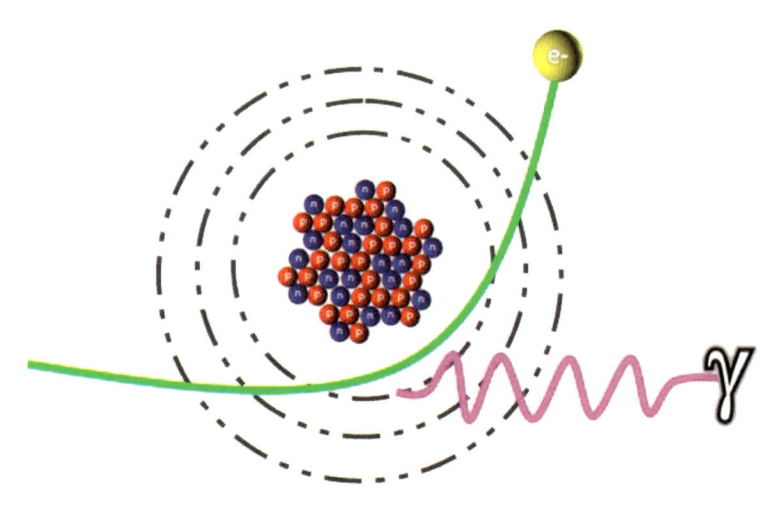

Figure 26 Bremsstrahlung

Bremsstrahlung is the energy emitted during deceleration of a charged particle. Bremsstrahlung is caused by a decrease in energy of a Beta particle from coming in close proximity to a heavy nucleus.

6.3 Recall factors that influence the attenuation of radiation in matter

Gamma radiation is the most difficult to shield (attenuate). For the same given thickness, lead would be the most effective for attenuating gamma radiation.

Water would be most effective material for attenuating Neutron radiation, assuming all materials are the same thickness.

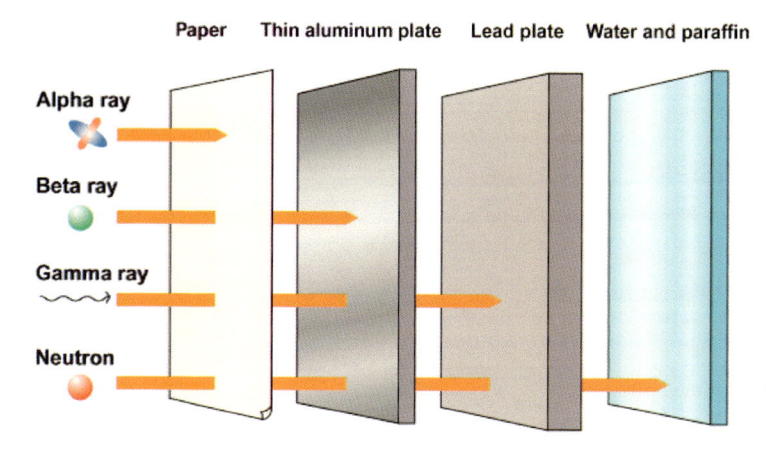

Figure 27 Radiation Penetration

Typical Shielding Characteristics
- **Alpha:** Thin amounts of most materials (paper, unbroken dead cell layer of skin, few cm of air)
- **Beta:** Low Z and low density materials (rubber, plastic, aluminum, glass)
- **Gamma:** High Z and high density materials (lead, steel, depleted Uranium, Tungsten)
- **Neutron:** Hydrogenous material for moderation (oil, poly plastic, water) and capture material for absorption (Boron, Cadmium)

6.4 Recognize the relationship of atomic number of the shielding material and its ability to attenuate Alpha or Beta radiation.

6.4.1 Alpha Shielding

Due to being the largest sized particle and having a plus two charge it is easiest to attenuate **Alpha** radiation.

A thin sheet of paper will usually be sufficient to stop most Alpha particles.

6.4.2 Beta Shielding

Because of its low atomic number, Aluminum is a better high energy Beta shield than Lead to minimize Bremsstrahlung Production.

6.5 Recall the values of Half Value Layers (HVL) or Tenth Value Layers (TVL) for Cobalt-60 (Co-60) gamma rays for lead, steel, concrete and water

Co-60 has a ~1 MeV gamma

1 MeV	HVL*	TVL*
Lead	.5"	1.5"
Steel	1"	3"
Concrete	4"	12"
Water	8"	24"

See section 19.2 for further information.

Pro Tip: An easy way to memorize the table above is that the TVL is 3x the HVL, so you only need to remember the HVL. The <u>half</u> value layer starts at <u>half</u> an inch. It is then doubled (x<u>2</u>) skipping <u>2</u>" after lines between steel and concrete. You can remember the order of Lead, Steel, Concrete and Water (LSCW) with a mnemonic **L**ittle **S**quirrels **C**ollect **W**alnuts.

6.6 Solve total dose problems given a dose rate or curie content values for various types of radiation

If the exposure rate at two feet from a point source is 20 R/hr, 5 R/hr is the expected exposure rate at four feet from the source. (Double the distance = quarter the dose, aka the **Inverse Square Law**.)

$$I_1(d_1)^2 = I_2(d_2)^2$$

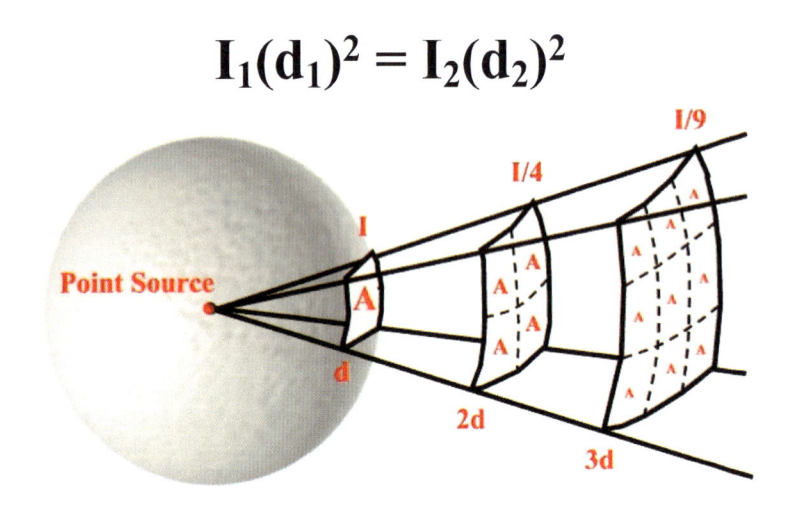

Figure 28 Inverse Square Law
Did you know that the inverse square law also effects gravity, light, and magnetic forces?

Practice:
A worker stands 12 ft. from a gamma point source reading 3 R/hr. at 3 ft. performing a task. The task requires the worker to remain in the same location for 3 hrs. We can use the $I_1(d_1)^2 = I_2(d_2)^2$ equation to find the answer, or we can notice that if you double the distance (3ft.) twice, you get 12 ft. So if you quarter the dose twice, you can find the dose rate is 187.5 mR/hr for 3 hours = 562.5 mRem is the approximate exposure received.

$$3ft \xrightarrow{1} 6ft \xrightarrow{2} 12ft$$

$$3000 \text{ mR/hr} \xrightarrow{1} 750 \xrightarrow{2} 187.5 \text{ mR/hr}$$

Or the long way:

$$\frac{I_1 \times (d_1)^2}{(d_2)^2} = I_2$$

$$\frac{3{,}000 \text{ } mR/hr \times 3ft^2}{12ft^2} = 187.5 \text{ } mR/hr$$

$$\frac{3{,}000 \times 9}{144} = 187.5 \text{ } mR/hr$$

Then multiply by the stay time:

$$187.5 \text{ } mR/hr \times 3 \text{ } hrs = 562.5 \text{ } mRem$$

Practice:
You have a point source containing **10 Ci of Cs-137**. You have a worker at **10 foot** for **10 minutes**. His total exposure would be **56.27 mRem**. Using $6CEN/d^2$, you can calculate that the exposure rate is 0.3376 R/hr. Convert that to 337.6 mR/hr., and then multiply it by his stay time. In this case, the say time is 1/6 of an hour which gives you a final answer of 56.27 mR total exposures.

$$\frac{6CEN}{d^2} \times Stay\ Time = Total\ Exposure$$

$$\frac{3.376 \times 10\ Ci}{10^2} = 0.3376\ R/hr$$

$$\frac{33.76}{100} = 0.3376\ R/hr$$

Or the long way:

$$\frac{6 \times 10\ Ci \times (0.662 \times 0.85)}{10^2} = 0.3376\ R/hr$$

Then divide by the stay time:

$$\frac{337.6\ mR/hr}{6} = 56.27\ mR/hr$$

Practice:
A survey of the job site indicates a dose rate of 360 mR/hr. The worker indicates the job will take 15 minutes with the wrench listed on the work order (90 mRem total exposure). The worker instead uses a wrench with a longer handle, tripling the worker's distance from the hot spot and increasing the worker's time in the area to 30 minutes. We need to figure out what the dose would be at 'triple the distance", at the new stay time. Since we don't know the exact distance, we can use 1 and 3 in our calculations for distance. We can use the $I_1(d_1)^2 = I_2(d_2)^2$ equation to find that (360 mR/hr. * 1^2)/3^2 = 40mR/hr. And that **20 mRem** is the worker's dose using the long handled tool.

$$\frac{I_1 \times (d_1)^2}{(d_2)^2} = I_2$$

$$I^2 \times Stay\ Time = Total\ Exposure$$

$$\frac{360\ mR/hr \times (1)^2}{(3)^2} = 40\ mR/hr$$

$$\frac{360\ mR/hr}{9} = 40\ mR/hr$$

$$40 mR/hr \times .5hr = 20\ mRem$$

Practice:

Now your boss wants to know how much dose was saved using the long handled tool for some meeting he has with the plant manager. 360 mR/hr. divided by 3^2 = 40 mR/ hr. at 3 times the distance, for a total dose of 20 mRem for the job, versus the 90 mRem it was going to take, **saving 70 mRem**. Way to go ALARA Master!

360mR/hr x .25 hours = 90 mRem

90 mRem − 20 mRem = 70 mRem Savings

Practice:
The exposure rate at 10 ft. from a small valve is 15 mR/hr. A mechanic will be working for 3 hours at a distance of two 2 ft. from this valve. We can use the $I_1(d_1)^2 = I_2(d_2)^2$ equation to find that 15 mR/hr. * 10^2 ft. / 2^2 ft. * 3 hours = **1,125 mR** is the mechanics expected total exposure for this job.

$$\frac{I_1 \times (d_1)^2}{(d_2)^2} = I_2$$

$$I^2 \times Stay\ Time = Total\ Exposure$$

$$\frac{15\ mR/hr \times (10)^2}{(2)^2} = 375\ mR/hr$$

$$\frac{15\ mR/hr \times 100}{4} = 375\ mR/hr$$

$$\frac{1,500\ mR/hr}{4} = 375\ mR/hr$$

$$375\frac{mR}{hr} \times 3hrs = 1,125\ mRem$$

6.7 Calculate the exposure rate for a specific radionuclide given the Gamma ray constant, distance from the source(s), and the source activity in Curies (Ci)

To determine the gamma exposure rate in R/hr. at one foot from a radioactive point source use the **6CEN** formula.

Where:
C = source activity in **C**uries (Ci)
E = Gamma **E**nergy in M**e**V
N = **N**umber (%) photon yield

Pro Tip: The predominant gamma emitters in nuclear power plants are Co-60 and Cs-137. Knowing this, you can make the equation simpler by calculating the 6 x E x N beforehand as they are constants, then multiply that number by the given Curie content.

82

Co-60 emits a 1.173 MeV & 1.332 MeV gamma 100% of the time.

$$(1.173 \times 100\%) + (1.332 \times 100\%) = 2.505$$

$$2.5 \times 6 = 15$$

Shortcut for 6CEN for Co-60 is: **15 x C**uries.

Cs-137 emits a 662 KeV (.662 MeV) gamma 85% of the time.

(0.662 x 0.85 = 0.563)

0.563 x 6 = 3.376. (About 3.4)

Shortcut for 6CEN for Cs-137 is: **3.38 x C**.

Practice:
You have a point source containing **5 Curies of Co-60**. You want to know what the exposure rate at **1 foot** is. Using 6CEN, you can calculate that the exposure rate is **75.15 R/hr**.

$$15 \times 5 \text{ Ci} = 75 \text{ R/hr}$$

Or the long way:

$$6 \times 5\text{Ci} \times ((1.173 \text{ MeV} \times 1) + (1.332 \text{ MeV} \times 1))$$

Practice:
You have a point source containing **10 mCi of Cs-137**. You want to know what the exposure rate at **1 foot** is. Using 6CEN, you can calculate that the exposure rate is **33.8 mR/hr**.

Since the Curies and the R/hr are the same units (milli) you don't need to convert them.

$$3.38 \times 10 \text{ mCi} = 33.8 \text{ mR/hr}$$

Or the long way:

$$6 \times 10 \text{ mCi} \times (0.662 \text{ MeV} \times 0.85) = 33.8 \text{ mR/hr}$$

To calculate the gamma exposure rate at different distances than 1 foot, you use the inverse square law to figure it out (Divide it by the distance squared).

$$\frac{6CEN}{d^2}$$

On the chalkboard:

$$6CEN/d^2$$

$$15 \times 2 / 9 = 3.33 \text{ R/hr}$$

The Long Way:
$$\frac{6 \times 2Ci \times (1.173 \text{ MeV} + 1.332 \text{ MeV})}{3 \times 3} = 3.33 \text{ R/hr}$$

Practice:
You have a point source containing **2 Curies of Co-60**. You want to know what the exposure rate at **3 foot** is. Using 6CEN/d², you can calculate that the exposure rate is **3.33 R/hr**.

$$\frac{15 \times 2\ Ci}{3^2} = 3.33\ R/hr$$

$$\frac{30\ Ci}{9} = 3.33\ R/hr$$

Or the long way:

$$\frac{6 \times 2\ Ci \times (1.173\ MeV + 1.332\ MeV)}{3^2} = 3.33\ R/hr$$

Practice:
You have a point source containing **10 mCi of Co-60**. You want to know what the exposure rate at **5 foot** is. Using 6CEN/d², you can calculate that the exposure rate is **6 mR/hr**.

$$\frac{15 \times 10\ mCi}{5^2} = 6\ mR/hr$$

$$\frac{150}{25} = 6\ mR/hr$$

Or the long way:

$$\frac{6 \times 10\ mCi \times (1.173\ MeV + 1.332\ MeV)}{5^2} = 6\ mR/hr$$

Practice:
You have a point source containing **1 Ci of Cs-137**. You want to know what the exposure rate at **2 foot** is. Using 6CEN/d², you can calculate that the exposure rate is **0.844 R/hr. (844 mR/hr.).**

$$\frac{3.378 \times 1\ Ci}{2^2} = 0.844\ R/hr$$

$$\frac{3.378}{4} = 0.844\ R/hr$$

Or the long way:

$$\frac{6 \times 1\ Ci \times (0.662\ MeV \times 0.85)}{2^2} = 0.844\ R/hr$$

7 Decay Modes, Decay Rates, Half-Lives, and the Curie

The NUF covers four objectives in this category, one of which will be a calculation.

A Curie is the amount of an element that decays at a rate of 3.7E10 disintegrations per <u>second</u>.

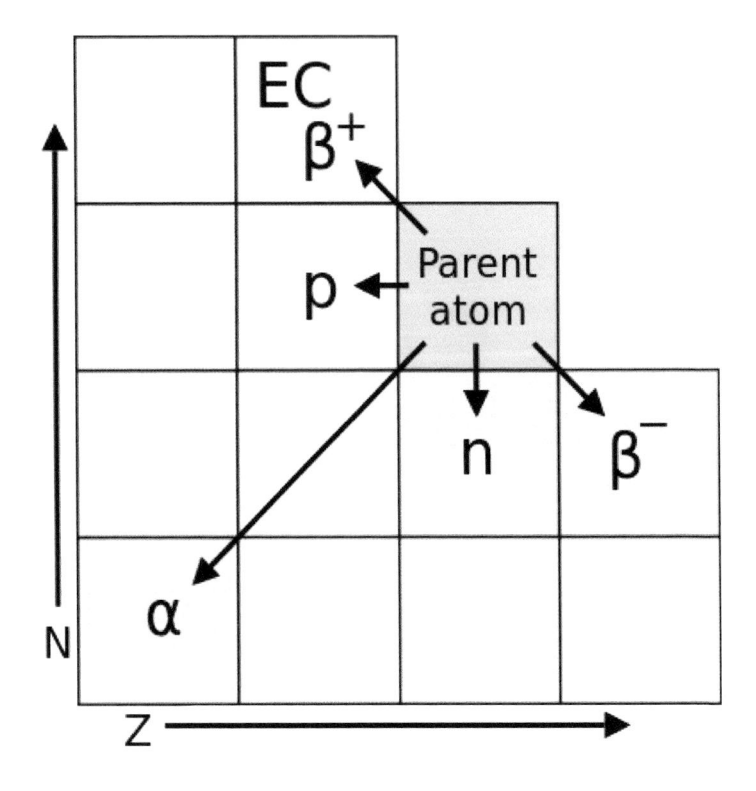

1 Ci = 3.7E10 DP<u>S</u> = 2.22E12 DP<u>M</u>

7.1 Define Half-Life

Half-Life is defined as the time span it takes for one half of the radioactive material in a sample to decay to something else.

- After 3 Half-Lives the original activity of a nuclide will be reduced by a factor of 8.
- After 7 Half-Lives less than 1% of the original activity will exist.

```
1 Half-life  = 1/2   (Factor of 2)   or 50%
2 Half-lives = 1/4   (Factor of 4)   or 25%
3 Half-lives = 1/8   (Factor of 8)   or 12.5%
4 Half-lives = 1/16  (Factor of 16)  or 6.25%
5 Half-lives = 1/32  (Factor of 32)  or 3.13%
6 Half-lives = 1/64  (Factor of 64)  or 1.56%
7 Half-lives = 1/128 (Factor of 128) or 0.78%
8 Half-lives = 1/256 (Factor of 256) or 0.39%
```

7.2 Describe the relationship between the radioactive decay constant and the Half-Life of a nuclide

The relationship of the decay constant and the Half-Life is described as: "Both remain constant and are closely related".

The mathematical expression which expresses the relationship between the radioactive **decay constant** and the Half-Life of a nuclide (to convert from Half-Life to decay constant or vice versa) is "$t_{1/2} = 0.693 / \lambda$". That is the decay formula and uses λ (Lambda) as the decay constant.

$$T_{1/2} = .693 / \lambda$$

The decay constant for a given radionuclide is equal to the natural log of 2 divided by the Half-Life of the radionuclide.

$$\lambda = \frac{\ln 2}{T_{1/2}}$$

7.3 Recognize the relationship between Effective Half-Life, Radiological Half-Life and Biological Half-Life

Effective Half-Life is defined as the time it takes for internally deposited radioactive material to be reduced to one-half of the initially deposited activity.

The Effective Half-Life (T_{eff}) of a radionuclide can be determined by multiplying the radioactive ($T_{1/2}$) and biological (T_b) Half-Life of the nuclide together and then dividing by the sum of the radioactive and biological Half-Life of the nuclide.

The Effective Half-Life is always <u>less</u> than either the Biological Half-Life or the Radioactive Half-Life.

$$T_{eff} = \frac{(T_{1/2} \times T_b)}{(T_{1/2} + T_b)}$$

Pro Tip: Remember that the multiplication is on top, addition is on bottom. Think of it as an arrow with the X being the fletching and the + being the tip of the arrow in the ground.

Practice:
For an internal deposit of Co-60, if the biological Half-Life is 8 days and the radiological Half-Life is 5.27 years, 7.97 days is the effective Half-Life.

$$T_{eff} = \frac{(T_{1/2} \times T_b)}{(T_{1/2} + T_b)}$$

$$T_{eff} = \frac{(5.27 \text{ years} \times 8 \text{ days})}{(5.27 \text{ years} + 8 \text{ days})}$$

$$T_{eff} = \frac{((5.27 \times 365) \times 8)}{((5.27 \times 365) + 8)}$$

$$T_{eff} = \frac{(1924 \times 8)}{(1924 + 8)}$$

$$7.97 \text{ days} = \frac{(15,392)}{(1,932)}$$

7.4 Solve or manipulate radioactive problems given the use of a calculator, the quantity of curies of a nuclide and its Half-Life

If asked to calculate the present activity of a source with an original activity of 200 Curies (Ci) and a Half-Life of 2.85 years. The formula you would use is:

$$A = A_o e^{-\lambda t}$$

$$A = 200 \text{ Ci } e^{-(0.693/2.85 \text{ yr})(t)}$$

Pro Tip: Another way to calculate Half-Life, which is faster and easier:

$$A = A_o \times 0.5^{t/t_{\frac{1}{2}}}$$

To get the current activity "A", you multiply the original activity "A_o" times 0.5 to the "n" power, where "n" is the elapsed time divided by the Half-Life.

Practice:
If given a radioactive sample containing 2.4 Curies with a 1.5 year Half-Life, using $A = A_o (.5)^{t/t_{1/2}}$ you would calculate the activity present in the sample after 10 months to be 1.63 Curies.

$$A = A_o \times 0.5^{t/t\frac{1}{2}}$$

$$A = 2.4Ci \times 0.5^{\frac{10mo}{1.5y \times 12mo}}$$

To figure how many half-lives have elapsed, we first convert years to months (1.5 years x 12 months = 18 months). Then we divide the elapsed time of 10 months by 18 months.

$$A = 2.4Ci \times 0.5^{0.556}$$

10 months is 0.556 Half-Lives for this material. We then raise '0.5' (half) to that value, to find out how much of the original material is left.

$$A = 2.4Ci \times 0.68$$

0.556 half-lives later, there is 68% (0.68) of the original material/activity left. We multiply that by the amount of the original material:

$$1.63Ci = 2.4Ci \times 0.68$$

Summary:
Level 3 Area: bg/a <300
- Alpha is Primary (>90%) hazard
- Count smears for alpha
- Count air samples for alpha
- Posted "Alpha Level III area"
- ≤ 50:1 Posted "**Alpha frisking/monitoring is required upon exit**"
- Equipment and materials exiting Level III areas should be properly labeled
- Segregate equipment and materials ≤ 50:1

Level 2 Area: bg/a= 300-30,000
- Alpha is Significant hazard, 10-90% of hazard is alpha, 50% @ 3000
- Count smears for alpha when the beta-gamma contamination > 20k dpm/100cm2
- Count air samples for alpha when the beta-gamma contamination >.3 DAC
- >100 dpm/100cm2 alpha sample results requires additional samples

Level 1 Area: bg/a >30,000
- Alpha is Minimal (<10%) hazard, primary hazard is beta-gamma
- Count smears for alpha when the beta-gamma contamination > 100k dpm/100cm2
- Count air samples for alpha when the beta-gamma contamination >1 DAC
- >100 dpm/100cm2 alpha sample results requires additional samples

8.1 Define beta-gamma to alpha ratio and how it is determined

Activity Ratio = beta-gamma / alpha

Pro Tip: Normally higher numbers mean higher hazards. However, a higher beta-gamma / alpha ratio means a lower hazard, and a low beta-gamma / alpha ratio is a higher hazard.

Practice:
A valve reads 150k dpm/100cm^2 (beta-gamma) and 50 dpm/100cm^2 alpha.

15,000/50 = 3,000 beta-gamma / alpha ratio.

8.2 Describe the action levels and controls for alpha monitoring using beta-gamma ratios, contamination survey data, and air sampling results

Level 3 Area: bg/a <300
- Alpha is Primary (>90%) hazard
- Count smears for alpha
- Count air samples for alpha
- Posted "Alpha Level III area"
- ≤ 50:1 Posted "**Alpha frisking/monitoring is required upon exit**"
- Material exiting Level III areas should be labeled
- Segregate equipment and materials ≤ 50:1

Level 2 Area: bg/a= 300-30,000
- Alpha is Significant hazard, 10-90% of hazard is alpha, 50% @ 3000
- Count smears for alpha when the beta-gamma contamination > 20k dpm/100cm^2
- Count air samples for alpha when the beta-gamma contamination >.3 DAC
- >100 dpm/100cm^2 alpha sample results requires additional samples

Level 1 Area: bg/a >30,000
- Alpha is Minimal (<10%) hazard, primary hazard is b-g
- Count smears for alpha when the beta-gamma contamination > 100k dpm/100cm^2
- Count air samples for alpha when the beta-gamma contamination >1 DAC
- >100 dpm/100cm^2 alpha requires additional samples

Practice:
You perform a survey in alpha Level 1 area. The general area beta-gamma contamination is 15k dpm/100cm^2. A valve reads 150k dpm/100cm^2 (beta-gamma). What is your action?

Action: In a Level 1 you would count smears for alpha when the beta-gamma contamination > 100k dpm/100cm^2., and calculate the beta-gamma to alpha ratio.

You count the valve smear for alpha; it is 50 dpm/100cm^2.

15,000/50 = 3,000 beta-gamma / alpha ratio.

Action: An Area with a beta-gamma / alpha ratio between 300 and 30,000 is Level 2 area. The area is re-posted as alpha Level 2 area.

8.3 Describe the benefits and limitations for each of the following individual monitoring techniques:

Personal Air Samplers (PAS)

Potential Dose	Level	Technique
> 10 mRem	Screening	Whole body counting, PAS, or excreta
> 100 mRem	Verification	Excreta measurements
> 500 mRem	Investigation	Extensive excreta measurements

- The results from PAS can be used to determine individual intake and dose from routine work activities.
- Whenever a PAS indicates a potential exposure may unexpectedly exceed the screening level of 10 mRem committed effective dose, action should be taken to confirm the extent of exposure.
- Where PAS results indicate potential exposures exceed the verification level of 100 mRem committed effective dose, excreta measurements should be used to investigate and determine the alpha intake.

Whole body counting (WBC)

WBC methods are limited for detection of alpha internal contamination because most alpha emitting radionuclides are not accompanied by gamma photon emissions with sufficient energy to be detected by whole body counting.

Excreta sampling (urine & feces)

Excreta is used to confirm the magnitude of the airborne activity intake.

Congratulations! You have reached the finish line. **None of the material after this is covered on the NUF exam**. However, you may want to review the meter reading test in section 10, most sites give this test now as well.

NukeWorker Online Practice Test

This book is accompanied by the online NUF RP practice test found at NukeWorker.com. It is highly recommended that you take the practice test many times. The practice test is comprised of 50 questions randomly generated from a large database of questions, and can cover any of the objectives. Due to the random nature of how the practice test is generated, it may not cover all the objectives, and it may cover an objective more than once so it is recommended that you take the practice test many times. The practice test only gives you 60 seconds to answer each question to simulate a little pressure.

Meter Reading Training

Many sites now give a timed "meter reading" exercise, where the technician is given a picture of a meter face and they are required to answer what the correct reading is. Sometimes the test asks for gross CPM answers for friskers, other times they ask for DPM with the assumption of 10% efficiency and a given background.

The needle is on the 320 mark, the meter is on the x10 scale (320 x10 = 3,200 CPM) or 32k DPM

The needle is on the 100 mark, the meter is on the x100 scale
(100 x100 = 10,000 CPM) or 100k DPM

The needle is on the 220 mark, the meter is on the x100 scale
(220 x100 = 22,000 CPM) or 220k DPM

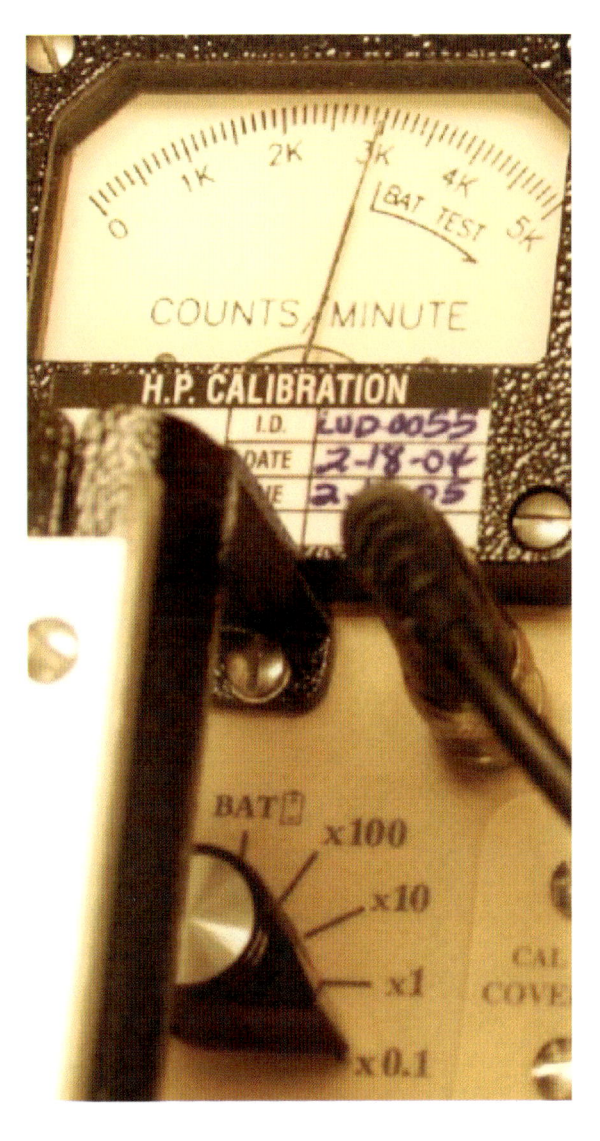

The needle is on the 3,000 mark, the meter is on the x0.1 scale (3,000 x0.1 = 300 CPM) or 3k DPM

The needle is on the 2,000 mark, the meter is on the x10 scale (2,000 x10 = 20,000 CPM) or 200k DPM

The needle is on the 4 mark, the meter is on the 500 mR/h scale, so the meter is reading 400 mR/hr

The needle is on the 3.6 mark, the meter is on the 5 mR/h scale, so the meter is reading 3.6 mR/hr

The needle is on the 2.5 mark, the meter is on the 5,000 mR/h scale, so the meter is reading 2,500 mR/hr or 2.5 R/hr

The needle is on the 0.85 mark, the meter is on the 50 R/h scale, so the meter is reading 8.5 R/hr

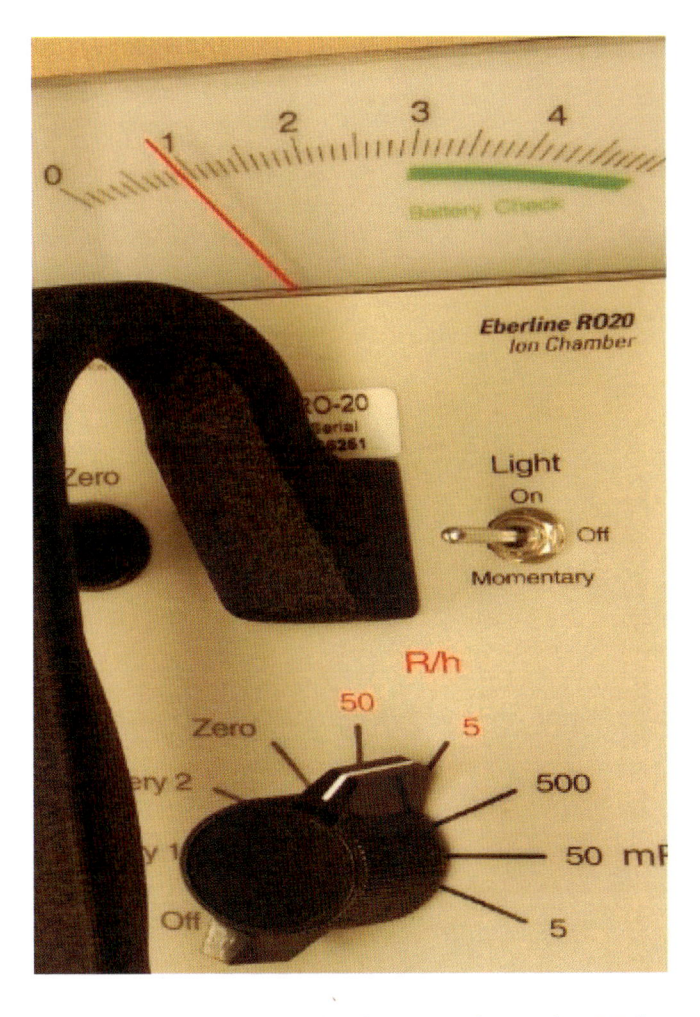

The needle is on the 0.9 mark, the meter is on the 5 R/h scale, so the meter is reading 0.9 R/hr or 900 mR/hr

The needle is on the 3.6 mark, the meter is on the 50 mR/h scale, so the meter is reading 36 mR/hr

Reference Material

11.1 Math/Physics/Chemistry

Work = Force x Distance
Power = Work / Time
$eV = \frac{1}{2}mv^2$
eV = charge x V
Electric Force = $k(Q_1 Q_2)/r^2$
Volts = Amps x Ohms
Charge = $1.6E^{-19}C$
tMw(.32) = Mwe
Inch = 2.54 cm
C = (F-32)/1.8
Ft^3 x 2.83E4 = mL or cc
1 Liter = 1000 cc

T = tera 10E12
G = giga 10E9
M = mega 10E6
k = kilo 10E3
m = milli 10E-3
u = micro 10E-6
n = nano 10E-9
p = pico 10E-12

1 σ = 68.27%
2 σ = 95.45%
3 σ = 99.73%

11.2 Nuclear Physics

p^1 = 938.25 MeV
n^o = 939.55 MeV
AMU = 931.48 MeV
Atomic Mass = Actual Atomic Mass (AMU)
33.9 eV/Ion Pair (air)
A = Z + n^o = Atomic Mass Number = Nucleons
Z = Protons = Atomic Number

Iso<u>mer</u> = same A & Z diff energy (<u>met</u>a)
Isoto<u>nes</u> = same # n^o
Isoto<u>pes</u> = same Z
Iso<u>bar</u> = same A diff Z (bar = weight)

λ = ln(2 or 10)/t½
n = t/t½
A = $A_o e^{-\lambda t}$ or **A = A_o(.5)^n** (for HVL or t½)
A = A_o(.1)^n for TVL
t = A/A_o(.693/t½)
A_o = A/(.5)^n
Line: A_o x D_o = A x D
Point: A_o x D_o^2 = A x D^2
TVL = HVL x 3 **NUF**
t½eff = (t½A **x** t½B)/(t½A + t½B)
(6)CEN/ d^2 ft (.5)CEN/ d^2 m
uCi/cc = ncpm/(eff)(vol cc)(2.22E6)

11.3 Dosimetry

1 Ci = 2.22E12 DPM = 3.7E10 Becquerel (1 DPS)
1 R = 87 ergs/gm air (~96 ergs/gm body)
Rad = 100 ergs/gm material
1 Gr**a**y = 100 R**a**d = 1J/Kg
Si**e**vert = 100 R**e**m
Rem = Rad x QF
Shallow Dose (skin) = 7 mg/cm^2 = 0.007 cm
Lens Eye = 300 mg/cm^2 = .3 cm
Intake = $\sum$ (IRF x A)/ $\sum$ (IRF^2)

11.4 Counting Statistics

2929 MDA = 3 + (4.65*$\sqrt{bg}$)/Eff
(10 min bg, 1 min sample)
2929 MDC = bg + 3 + (4.65*$\sqrt{bg}$)
(10 min bg, 1 min sample)

11.5 Background Radiation

Natural Radiation:
Terrestrial 28
Cosmic 27
Internal 39
Radon 200

Man-made Radiation:
Fallout <1
Nuke Facilities <1
Med 53
Consumer 10

11.6 Regulatory Limits & Postings

Post Airborne @ .3 DAC **NUF** or .1 DAC **DOE**
Deep Dose (WB) = 1000 mg/cm^2 = 1 cm
1 ALI = 5 Rem CEDE or 50 Rem CDE = 2000 DAC hrs
1 DAC = 2.5 mRem
DDE (WB) = 5 Rem/yr
CDE (organ) = H_{50} = DDE x 10 = 50 Rem/yr
TEDE = CEDE + DDE
CEDE = H_{50} x W_T = 5 Rem x I/ALI
LDE = DDE x 3 = 15 Rem/yr Life
Extremity = DDE x 10
Embryo/fetus = 500 mRem/gestation
PSE = TEDE/yr, 5 X TEDE/
Minors/Public = 100 mR/yr (10%)

2 mR/hr or 25 mRem/yr = Restricted Area
5 mR/hr @ 30 cm = Radiation Area
100 mR/hr @ 30 cm = High Radiation Area
500 Rads/hr @ 1 m = Grave Danger, Very High Radiation Area

11.7 ICRP Publications

http://www.icrp.org/publications.asp

ICRP 23: Reference Man
ICRP 30: Internal Dose
ICRP 60: 10CFR20/835

NukeWorker Comic:

NukeWorker Comic:

About The NUF

The Nuclear Utilities HP/RP Technician Fundamentals Exam (NUF) is a test given via the NANTeL system, and proctored and accepted at (most) all commercial nuclear power plants in the United States. It replaced the North East Utilities (NEU) Radiation Protection (RP) Exam. The results from the NUF are stored in NANTeL, as well as in the shared Personnel Access Data System (PADS).

NUF Question Database

The objectives and question database is hosted by the Institute of Nuclear Power Operations (INPO) in the NANTeL system. It is under the guidance of the 'NUF Users groups and their steering committee. It is proceduralized by the Nuclear Energy Institute (NEI). The governing document is NEI procedure NEI 03-04 "Guideline for Plant Access Training", specifically Appendix C, which is titled Section 11, "Health Physics Contractor Examination". The most recent version as of this writing was Revision 9, December 2014.

There are 52 objectives that the NUF is based on. The NUF has one question randomly chosen for each of the 52 objectives, for a total of 52 questions. Each objective draws a question at random from a databank of questions that relates to its objective. You will only receive one question that pertains to each objective, but you WILL receive one question that pertains to each of the 52 examinations' objectives. This book lists and covers those 52 objectives. Additionally, the order of the questions is random, so you most likely will NOT get a question from objective one first, then a question from objective two next. You may get a question from objective 30 first, then four next, etc.

There are 10 objectives under the heading of "Radiation & Contamination Sources", so of the 52 questions on your test, there will be 10 questions about Fission and Activation products. There are 14 objectives under the heading of Radiation Detection & Instrumentation, so there will be 14 questions about instruments. That means about half of the 52 questions are about instruments, and activation/fission products. Pay special attention to those areas.

Taking the NUF Exam

The test is multiple choice, there is no time limit. You may review and change any answer prior to submitting the test for grading. Once graded, you are able to review the questions again and to see the correct answer. You are allowed to bring a calculator, a pen or pencil, and some scratch paper. The NANTeL system allows you to attempt the test twice.

NUF Security

- No written material, including scrap paper, may be removed from the test area. This includes electronic notes.
- You are not allowed to bring any notes or other reference material.
- The exam must be proctored by a qualified proctor.

NUF Requirements

Contract Senior HP/RP techs must take the Nuclear Utilities HP/RP Fundamentals Exam (NUF) and pass with a grade of 80% or better to work at a commercial power plant in the US. Most plants require you recertify every 5 years, some require recertification every 4 years. Below are lists of how often you have to recertify as various sites. Additionally you cannot have had a break in the nuclear industry longer than 1 ½ years. You *typically* do not need to take the NUF if you are either an NRRPT "active participant" or have passed Part 1 of the ABHP CHP examination, however not *all* nuclear plant accept the NRRPT (we are told Seabrook does not). NUF, NRRPT and CHP certifications must be verified so bring a copy of your certification with you. NUF is typically saved in PADS, but not always so know where and when you last passed the NUF, in case it is not recorded in PADS.

Some sites require juniors to take the NUF as well.

5 Year NUF Sites:
Arkansas Nuclear One
Beaver Valley
Browns Ferry
Brunswick
Callaway
Columbia Generating Station
Comanche Peak
Cooper
Crystal River
D.C. Cook
Davis-Besse
Duane Arnold
Farley
Fermi
Fitzpatrick
Fort Calhoun
Grand Gulf
Hatch
H.B. Robinson
Indian Point
Kewaunee
Millstone
Monticello
North Anna
Palisades
Palo Verde
Perry
Pilgrim
Point Beach
Prairie Island
River Bend
Saint Lucie
Seabrook
Sequoyah
Shearon Harris
South Texas Project
Surry
Susquehanna
Turkey Point
V.C. Summer
Vermont Yankee
Vogtle
Waterford
Watts Bar
Wolf Creek

4 Year NUF Sites:
Braidwood
Byron
Calvert Cliffs
Catawba
Clinton
Dresden
Hope Creek
LaSalle
Limerick
McGuire
Nine Mile Point
Oconee
Oyster Creek
Peach Bottom
Quad Cities
R.E. Ginna
Salem
Three Mile Island

2 Year NUF Sites:
San Onofre

Non- NUF Sites:
Diablo Canyon

Resume Tips

15 Skills Power Plants Look for on Contract RP Resumes

When RP departments look at the past job history on your resume prior to an outage, they are looking for the following 15 skills defined in NEI 03-04 Appendix D, so that they can give you qualifications based on your prior work history. If these skills are not specifically listed on your resume, you may not be granted the qualification for that skill. Of course, don't put anything on your resume that you haven't done, but don't forget to take credit for what you have done.

1. Operation of Survey Instruments, and Count Rate Meters
2. Performance of Radiation and Contamination Surveys
3. Performance of Airborne Radioactivity Surveys
4. Radioactive Material Movement and Storage
5. Radiological Posting/De-posting
6. Response to Radiological Alarms
7. Operation Continuous Air Monitors
8. Provided Radiological Job Coverage
9. Provided High Risk Job Coverage
   (e.g. >1R/hr, >500k dpm/100cm2, >5 DAC)
10. Directed/Performed Area & Equipment Decontamination
11. Surveyed Material for Unconditional Release
12. Performance of Personnel Decontamination
13. Monitored & Coached Workers in the RCA Including Their Ingress & Egress
14. Operated HEPA Vacuum and/or Ventilation Equipment
15. Performed Remote Radiological Monitoring

An example of how it might look on your resume:

Start Date – End Date
Plant Name, Plant Location (Contract Company)
Senior Radiological Protection Technician
Operated survey instruments, count rate meters, Continuous Air Monitors, as well as HEPA equipment. Performed radiation and contamination surveys, unconditional release surveys, airborne radioactivity surveys, radiological posting/de-posting, radiological job coverage to include high risk job coverage of steam generator work (>1R/hr., >500k DPM/100cm2), remote radiological monitoring, area and equipment decontamination, as well as personnel decontamination. Supported radioactive material movement, responded to radiological alarms and monitored and coached workers in the RCA including their Ingress and Egress

Acceptable Experience & Training for HP/RP Techs

ANSI 18.1 & ANSI 3.1
Nuclear power plants are committed to using the criteria in one of the versions of the American National Standard for Selection and Training of Nuclear Power Plant Personnel. Either ANSI N18.1-1971 or its replacement, ANSI/ANS 3.1-1978. ANSI/ANS 3.1 was revised in 1981 and again in 1987. The 1987 version of ANSI/ANS 3.1 has been adopted as most commercial power plants revised their training programs.

ANSI N18.1-1971 (Section 4.5.2) states:
Technicians in responsible (senior) positions shall have a minimum of **two years** of working experience in their specialty. These personnel should have a minimum of one year of related technical training in addition to their experience.

ANSI/ANS 3.1-1978 (Section 4.5.2) states:
Technicians shall have **three years** of working experience in their specialty of which one year should be related technical training.

ANSI/ANS 3.1-1987 (Section 4.5.3.2) states:
- Education:
    o High School Diploma
- Experience:
    o Radiation Protection: 2 years
    o Nuclear Power Plant:1 years
    o On site: 3 months
- Training:
    o As specified in Section 6

The "Section 6" criteria covers initial and continuing training after an individual is employed as a technician but not pre-employment training.

Definition of a "Year"

A maximum of 50 hours per week is accepted as counting towards experience. Stated another way, a "year" consists of 2000 hours worked in no less than 40 weeks. With this definition, in order to have two years of "experience," a technician would need at least 80 weeks of employment and a documented 4000 hours of acceptable health physics experience to be a two year senior HP/RP (120 weeks/6000 hours for three year Senior).

Acceptable Training for HP/RP Techs

Training	Credit
Certificate Program in Rad Protection	9 months
AS in Radiation Protection	1 year
AS in Rad Health	1 year
AS/advanced degree in Rad Health	1 year
AS/advanced degree in science/engineer	1 year
Navy Engineering Laboratory Tech	1 year
Non-university health physics courses	Case-by-Case
Utility-sponsored training programs	< Year

Acceptable Experience for HP/RP Techs

Job/Experience Type	Credit
Navy ELT (non-overhaul)	1:1 ≤ 1y
Navy ELT (overhaul)	1:1 NL
Shipyard/Tender RadCon	1:1 NL
National Laboratory	1:1 NL
Fuel Reprocessing/Production	1:1 NL
NPP Sr. or Jr. HP Tech	1:1 NL
NPP Dosimetry Tech	1:1 ≤ 6m
NPP Respiratory Prot. Tech	1:1 ≤ 6m
NPP Count Room Tech	1:1 ≤ 6m
NPP Control Point Monitor	1:1 ≤ 3m
NPP Laundry Monitor	1:1 ≤ 3m
NPP Decon (with surveys)	1:1 ≤ 3m
NPP GET HP Instructor	1:1 ≤ 6m
NPP HP Tech Instructor	1:1 ≤ 1y
Radioactive Facility D&D	CBC
Misc. HP Work at Other Sites	CBC

Each jobs' duties are reviewed to confirm the acceptability of the experience. 12 months of HP job coverage experience is required for qualification as a Sr. technician.

Old NEU RP Exam

Prior to Feb. 2004, the most widely accepted standardized RP Fundamentals exam for Contractor HP/RP Technicians was the Northeast Utilities HP/RP Tech Exam (NEU). The NEU had previously been accepted at many but not all commercial nuclear power plants in the United States. It originated, and was administered by the Northeast Utilities, owner of the Millstone Nuclear Station. The primary point of contact was a man named James Bennett, from the Millstone Nuclear Station who was one of the founding members of the 'steering committee' of the NUF users group. The NUF Users Group was created because the Northeast Utilities (NU) sold Millstone and indicated it was leaving the Nuclear Power Generation industry.

The original NEU exam series were paper tests covering the required objectives of that time. Each test was named HP Contractor #1, HP Contractor #2, HP Contractor #3, etc. There were roughly 22 different versions of the paper tests generated over the years. Those tests were not random. There were a limited number of tests generated each year, so you had a 20% chance to get the exact same test your friend took. The paper test was not easily or quickly changed. This resulted in compromises to the NEU test.

Quote from RDTroja on NukeWorker.com:
Many moons ago, when reactors were made of wood and half-lives were measured with sundials there were no standardized tests... in fact in some plants there were no tests at all. After a while, most of the utilities (and even some plants within the same utility) began developing their own tests to validate contract HPs' resumes... largely due to the fact that a great deal of them were more fiction than fact... and there was chaos in the world. Some plants' tests were easy and some were ridiculously hard. Some tested theory and some practical knowledge. Sometimes you could even get different answers to the same questions depending on what plant you went to. One day a now-defunct (merged into a larger group) utility known as North-East Utilities developed a test that seemed to be reasonably good and (at least at first) had some reasonable security behind it... and many of the other utilities began accepting the test as valid, thus relieving them of the burden of developing their own. Naturally the burden had to go somewhere, and it went to Northeast Utilities who eventually got tired of carrying the load for everyone else and abandoned the test. INPO liked the idea of a standardized test and they 'took up the mantle' so to speak and the test transmuted from the Northeast Utilities (NEU) Radiation Protection Exam to the Nuclear Utilities Fundamentals Radiation Protection Exam (NUF). And all was right with the world.

NUF Governing Bodies

NANTeL

National Academy for Nuclear Training e-Learning (NANTeL) is a national Web-based system that provides standardized nuclear power plant training courses for the supplemental workforce to any computer with internet access. The courses are required before workers can perform most jobs at a nuclear power plant. Training topics include plant access, radiation worker, and human performance tools. Institute of Nuclear Power Operations (INPO) and the National Academy for Nuclear Training manage and operate the system for the U.S. nuclear industry.

Currently, your user name for NANTeL is your Last name, followed by your first and middle initials, then your two digit month and day of birth.

INPO

The Institute of Nuclear Power Operations (INPO), established a comprehensive system of personnel training and qualification. It created the National Academy for Nuclear Training in 1985 to integrate the training programs of INPO, the training efforts of all U.S. nuclear energy companies and the independent activities of the National Nuclear Accrediting Board.

PADS

Through a series of buyouts and mergers, CANBERRA is the Nuclear Measurement Business Unit of AREVA, Inc. who administers the vital data sharing platform used by all Nuclear Power Plants in the US named the Personnel Access Data System (PADS).

For over 20 years, the NMBU has administered the Personnel Access Data System (PADS) system for the US Nuclear Power Plant industry. PADS is a network of computer systems and an associated central database designed to increase the efficiency of in-processing of workers into nuclear power plants through data sharing.

Information contained within PADS includes:

- Dosimetry
- Access data, including employee background check confirmations
- Fitness for duty information
- Training and skill-set data

The PADS system was initiated in the early 1980's by a group of utilities interested in expediting the process of staffing nuclear plants. A set of rules was agreed upon in 1988 and PADS was live by 1989. In 1994, the Nuclear Energy Institute (NEI) assumed oversight of PADS and has implemented the program with every operating nuclear power plant in the US. NEI 03-01, Nuclear Power Plant Access Authorization Program governs its operation.

PADS has been the software system used to track nuclear power plant workers for the entire US Nuclear Power Plant industry since 1999. Over 400,000 individual workers are documented in PADS.

Other RP Tests Available

NRRPT

Most plants waive the NUF requirement if you have an active NRRPT certification. The NRRPT was established in 1976 through the sponsorship of the Health Physics Society and the American Board of Health Physics.

The NRRPT has established a credentialing exam. This 150 question exam covers broad-based radiation protection knowledge of accelerators, university health physics programs, medical health physics, power reactors, government radiological facilities, radioactive waste disposal, and transportation of radioactive material, fundamentals, and regulatory requirements.

The NRRPT also provides incentives and services to encourage personnel to maintain and expand radiation protection education and training. The NRRPT has been endorsed in various ways by several organizations. The Institute for Nuclear Power Operations (INPO) and the Department of Energy (DOE) has openly recommended that nuclear facilities encourage their personnel to seek NRRPT Registration. The Nuclear Regulatory Commission (NRC) provides support by having a staff member assigned to the NRRPT Panel of Examiners.

Nuclear facilities (i.e., power plants, government facilities, universities, medical facilities, and military services) provide incentives for personnel to seek and maintain registration. Although registration does not constitute licensing nor does it guarantee the adequacy of an individual's performance, registration does test competency in fundamentals and operational topics. Registration has been established as a recognized mark of motivation and achievement of radiation protection personnel. The professional credential provided by the Registry has clearly stimulated interest in radiation protection training programs. Some companies specifically require registration for hiring, promotions, and salary grade increases. Some commercial power plant contractors' pre-employment screening exams have been waived for NRRPT Registered personnel.

ABHP CHP

The American Board of Health Physics is the certification body for the practice of professional health physics. The ABHP has existed since 1958 and is responsible for determining the qualifications of a Certified Health Physicist (CHP). Requirements to become a CHP are found on the main website and include academic and experience criteria as well as successful performance on two certification examinations. The Part I exam is in multiple choice format and covers fundamentals of health **physics**. The Part II exam is a combination of longer questions that cover a broader topic range. Members of the ABHP are appointed by the American Academy of Health Physics to 5-year terms. The Board members appoint members of two Panels of Examiners which prepare and grade their respective certification exams.

Issues with NUF 'Facts'

Roentgen vs Rem
In the context of this book mR or Roentgen and mRem and Roentgen Equivalent Man (Rem) are used interchangeably because that's how it is used on the NUF test. If you are curious; 1 Roentgen is 87 ergs/g in air for photons greater than 10 keV and less than 3 MeV, which is about 92 to 96 ergs/g in people (depends on if you are talking about bone, muscle or fat). The Rem is 100 ergs/gram in human tissue for photons of any energy. So 1 R Gamma is roughly 0.96 Rem Gamma.

Shielding
For purposes of the NUF, use the values in section 6.5. They are not 'correct' in the real world, but they are the numbers used in the NUF Study guide, and the NUF test. So, if you want to get the answers correct on the NUF test, use those numbers.

The NUF folks seem to have confused cm with inches. Additionally, the NUF uses the Half Value Layer x 3 as a Tenth Value Layer which is 'close' but also incorrect.

Real HVL 1 MeV Photon		w/Buildup
Lead	0.86 cm	1.31 cm
Steel/Iron	1.5 cm	3.45 cm
Concrete	4.5 cm	12.05 cm
Water	9.8 cm	28.71 cm

Isotope vs. nuclide
The NUF material often uses the word Isotope when they should use the term Nuclide.

Post Airborne

The NUF material, and many power plants post at 0.3 DAC, while most DOE facilities post at 0.1 DAC. This isn't actually an issue, just a fact to note for DOE technicians heading to the NRC arena.

Thermal Neutron Quality Factor (Q)

The NUF uses 2 for a quality factor for thermal neutrons. However the REAL number for "thermal" neutrons in the USA is 3, although it's 5 in most other countries. Technically, it's variable:

Neutron energy (MeV)	Quality Factor (Q)
2.5×10^{-8}	2
1×10^{-3}	2
1×10^{-2}	2.5
1×10^{-1}	7.5
5×10^{-1}	11
1	11
2.5	9
5	8
7	7
10	6.5
14	7.5
20	8
40	7
60	5.5
1×10^{2}	4
2×10^{2}	3.5
3×10^{2}	3.5
4×10^{2}	3.5

Quality factors are recommended by the ICRP, the NCRP, and the ICRU. But for regulatory purposes, one must use the factors of the NRC or the DOE, or other regulator.

History of NukeWorker.com

NukeWorker was founded on April 18th, 1999 by Michael D. Rennhack, a Radiological Engineer who has been in the nuclear business for more than a quarter of a century.

Michael had always been fond of computers, starting in 1982 at the age of 11 with his Commodore 64. Michael quickly modified the C-64 and used it as an electronic bulletin board system (BBS), where other people in town would call his computer with theirs and send email or exchange messages on the bulletin board. He ran this electronic community, known as "The Great Wall" until his nuclear career took him on the road in 1992, at which time The Great Wall fell.

Three years later, Windows 95 and the Pentium chip were released. 1995 was also the year that internet explorer and the Internet gained popularity. Michael purchased one of these new machines and started learning how it worked. In January of 1997, Michael produced his first website, a vanity site named Rennhack.net, which was composed of small pages containing information about his interests, which at the time was collecting Zippo lighters, Martial Arts, and his nuclear career. He monitored the popularity of each page, and noticed that the nuclear sections were getting all of the attention, so he focused on improving that portion of his site.

Over the next two years, Michael continued his work on the nuclear section of his site until it out grew Rennhack.net and needed a new home. So Michael purchased the domain name NukeWorker.com, which was inspired by the AOL user name of a fellow nuclear worker named Bob Reece (R.I.P.).

NukeWorker.com went online April 18th, 1999 with the mission of enriching nuclear workers lives and instantly gained a following of over 200 people the first day, and over 6,000 unique visits a month. The number of visitors has been growing ever since. In 2012, NukeWorker.com received more than half a million unique visits a month.

In 2002, the server bills for NukeWorker.com had increased by more than tenfold, and had become a financial burden for Michael who was offering all of the sites services for free while still working as a nuclear worker. The decision was made to allow people to help with the bills. NukeWorker was established as a corporation, and started accepting fees for advertising, outage schedules and job postings. Eighty percent of NukeWorker's bills were paid for by those donations and fees which were at one time free services. The other 20 percent was paid out of Michael's pocket. In 2004 NukeWorker established a "Nuclear Pride Shop" that sold shirts, hats, and stickers which helped with another 10% of the bills and started a grass roots marketing program. In 2006 online OSHA training was added by popular demand.

From its birth in 1999 to a fateful day in 2004, when Michael Rennhack was asked "What's NukeWorker?" the response to that question was that NukeWorker.com was a community for nuclear workers, expounding on the 7,300 pictures in the photo gallery, thousands of questions in the free online practice tests that had been used more than a million times, the free study guides, the 158,000 messages in the message board, the thousands of jobs in the job board, the nuclear news section that was updated hourly, the nuclear pride store, and the massive facility information section. These features made NukeWorker.com the most popular destination on the internet for nuclear workers. This complex answer made it hard for some companies to grasp how NukeWorker could help them.

However, if you asked anyone other than its founder that question, they would tell you simply that NukeWorker was a 'job board', a favorite destination for nuclear employers to find experienced and qualified nuclear workers. Michael heard a friend describe NukeWorker as "a job board" to a company at a local Health Physics Society meeting. He noticed that that simple answer was easy for the potential client to grasp.

That night, Michael went home and analyzed the websites income, and noticed that the job board was responsible for the majority of the sites revenue. Michael then realized what everyone else already knew. NukeWorker.com was indeed a job board for nuclear workers.

The site was adjusted to make it easier for employers to post jobs in late 2005, and was marketed as the nuclear job site it really was.

The first day NukeWorker went online in 1999, it had a simple logo; just a magenta tri-foil with the text 'NukeWorker' over top.

In 2001 the 'classic' orange and green NukeWorker character in a bubble suit holding a can with the nuclear symbol on it was born. It was the birth of the NukeWorker character.

In 2003 the NukeWorker logo underwent a transformation from the classic orange NukeWorker to the blue NukeWorker with a gray circular background.

To celebrate its 10 year anniversary in 2009, NukeWorker updated its logo again. The biggest change was to give the NukeWorker a friendly smile.

About the Author

Michael Rennhack started his nuclear career in 1989, founded NukeWorker.com ten years later in 1999 and has served as the President and Chief Executive Officer since its inception. Under his leadership, NukeWorker.com has experienced phenomenal growth and has become the nation's leading provider of nuclear job site services and the nuclear industry's most popular community.

Mr. Rennhack has been a member of the local and national Health Physics Societies since 1996. In 2000, Brookhaven National Lab honored Mr. Rennhack for his participation in their Ambassador Program for his work with the Boy Scouts, helping them earn their Atomic Energy Merit Badge. In 2008, Michael was elected Council Member of the East Tennessee Chapter of the Health Physics Society. In addition to founding NukeWorker.com, Mr. Rennhack has held radiological engineering, radiological management, and project management positions, and has worked at more than 50 nuclear facilities around the country.

In 2014 Mr. Rennhack was nominated for the prestigious Charles D. (Bama) McKnight Memorial Award, which is awarded by the NRRPT Board of Directors to persons who have given outstanding efforts in the radiation protection training field leading to increased knowledge and professionalism among Radiation Protection Technologists.

Born February 14th, 1971 in Michigan to Jerry and Betty Rennhack, Michael's love of computers began at age 11 when he programmed his first electronic community (BBS). Michael was the youngest board member of the local computer users group at age 14, and at age 18 he began his nuclear career as a decontamination technician. In 1999, Michael merged his passion for computers and the nuclear industry with the creation of NukeWorker.com, the first website dedicated to enriching nuclear workers lives.

Michael met his wife Kathy in February, 2002 while working in Minnesota and proposed to her in September of that year while vacationing in Paris. They were married in June, 2003 in Jamaica and promptly moved to East Tennessee. In April of 2005 they were blessed with their daughter Lisa then again in February of 2009 with a son, Thomas. In August of 2008 the Rennhack's returned to Michael's home town in Southwest Michigan.